T0189648

Topics in Intelligent Engineering and Informatics

Volume 7

Series Editors

J. Fodor, Budapest, Hungary
I. J. Rudas, Budapest, Hungary

For further volumes:
http://www.springer.com/series/10188

Gabriella Bognár · Tibor Tóth
Editors

Applied Information Science, Engineering and Technology

Selected Topics from the Field of Production Information Engineering and IT for Manufacturing: Theory and Practice

Editors
Gabriella Bognár
Tibor Tóth
Faculty of Mechanical Engineering and Informatics
University of Miskolc
Miskolc-Egyetemváros
Hungary

ISSN 2193-9411 ISSN 2193-942X (electronic)
ISBN 978-3-319-35036-3 ISBN 978-3-319-01919-2 (eBook)
DOI 10.1007/978-3-319-01919-2
Springer Cham Heidelberg New York Dordrecht London

Printed on acid-free paper

Springer is part of Springer Science+Business Media (www.springer.com)

Preface

The objective of the book is to give a selection from the papers that summarize several important results obtained within the framework of the József Hatvany Doctoral School operating at the University of Miskolc, Hungary. In accordance with the three main research areas of the Doctoral School established for Information Science, Engineering and Technology, the papers can be classified into three groups. They are as follows: (1) Applied Computational Science; (2) Production Information Engineering (IT for Manufacturing included); (3) Material Stream Systems and IT for Logistics. The volume contains 12 papers in all.

As regards the first area, some papers deal with special issues of algorithm theory and its applications, with computing algorithms for engineering tasks, as well as certain issues of database systems and knowledge intensive systems. In the following we attempt to give a brief summary of each paper.

Csendes, Cs. and Fegyverneki S.: *Parameter Estimation for Symmetric Stable Distribution by Probability Integral Transformation*. In this paper a new parameter estimation method for symmetric stable distributions is presented, which is a variation of maximum likelihood type M-estimators. A simulation study is performed to compare the proposed estimator with other methods based on performance properties and assessing convergence of the estimators.

Johanyák, Zs. Cs. and Kovács, Sz.: *Prediction of the Network Administration Course Results Based on Fuzzy Inference*. In the paper, the authors report on the development of a fuzzy model that is based on the previous performance of currently enrolled students, and gives a prediction for the number of students who will fail the exams of the Network Administration course at the end of the autumn semester (Least Squares-based Fuzzy Rule Interpolation).

Barabás, P. and Kovács, L.: *Optimization Tasks in the Conversion of Natural Language Texts into Function Calls*. Natural language processing (NLP) is a well known and an increasingly more important area in human–computer interaction. The goal of this paper is to develop a natural language framework, which will be used to extend existing systems with a natural language controlling capability.

Tóth, Zs. and Kovács, L.: *Pattern Distillation in Grammar Induction Methods*. The rule extraction phase plays a very important role in Context-Free Grammar induction systems. In the paper two novel methods are presented for pattern mining. The first is based on extended regular expressions and a multiplicity approach. The second method is based on the theory of concept lattices.

Mileff, P. and Dudra, J.: *Advanced 2D Rasterization on Modern CPUs*. This paper aims to investigate how effectively multi-core architecture can be applied in the two-dimensional rasterization process and what the benefits and bottlenecks of this rendering are. The authors answer the question of whether it would be possible to design a software rendering engine to meet the requirements of today's computer games.

Hriczó, K. and Bognár, G.: *Numerical Analysis of Free Convection from a Vertical Surface Embedded in a Porous Medium*. In this paper, the numerical solutions for free convective heat transfer in a viscous fluid flow over a vertical flat plate embedded in a porous medium under mixed thermal boundary conditions are examined. Applying a similarity transformation the transformed system of ordinary differential equations is investigated numerically.

Related to the second research area, except for the last paper dealing with a special measuring method, the focus is on Production Information Engineering with special regard to discrete production processes. As regards the Unhauzer paper, the author describes a new, complex IT solution suitable for measuring the relevant flicker parameters. In the following we give a brief summary of the papers.

Bikfalvi, P., Erdélyi, F., Kulcsár, Gy., Tóth, T. and Kulcsárné Forrai, M.: *On Some Functions of the MES Applications Supporting Production Operations Management*. As is known, the model-based decision support functions of business and manufacturing processes can be classified into different hierarchical levels, in accordance with their functions, objects, and time horizons. In this paper two methods for improving the quality of Production Operations Management (POM) are presented: a proactive one, using simulation-based fine scheduling, and a reactive one, based on evaluation of some Key Performance Indices (KPIs) determined from an efficient analysis of shop-floor production data. Both methods exploit the advantages of software applications used in different Manufacturing Execution System (MES) components.

Dudás, L.: *New Theory and Application for Generating Enveloping Surfaces without Undercuts*. The design and improvement of kinematical motion transfer surfaces (gear surfaces) require the modeling of surface–surface enveloping process and visualization of contact characteristics. To analyze the quality of mesh in respect of undercut, this study uses the special visualization capability of the Surface Constructor (SC) system developed by the author. The paper shows not only the theoretical background of the software but also gives a brief summary of a practical application.

Paniti, I.: *New Solutions in Online Sheet Thickness Measurements in Incremental Sheet Forming*. This paper discusses an approved analytical framework of Single Point Incremental Forming (SPIF) of sheet metals, which is capable of modeling the state of stress in the small localized deformation zone in case of corners, flat, and rotationally symmetric surfaces. The discussion focuses on the investigation of the sheet thickness prediction in the shell element used in the framework. Novel solutions are introduced in terms of online sheet thickness measurement and adaptive control in SPIF. A brief summary of a recently patented solution in Incremental Sheet Forming is also given in the paper.

Unhauzer, A.: *New Online Flicker Measuring Method and Module*. Flicker is a sequence of flashing lamp pulses which imperceptibly influences the human body and the environment. This paper describes a new online flicker measurement method and a multiple-tested software module based on multithreading technology that has been developed for the objective and exact analysis of electrical networks. The necessary theoretical background and the main development steps will also be shown.

The papers connecting with the third research field deal with different issues of materials stream systems and logistics. In the following we outline their content briefly.

Skapinyecz, R. and Illés, B.: *Presenting a Logistic Oriented Research Project in the Field of E-marketplace Integrated Virtual Enterprises*. The aim of the paper is to present an ongoing research project in the field of Virtual Enterprises, focusing mainly on perspectives of logistics (with special regard to freight transport) and e-commerce. In addition, the paper also gives a practical overview of the utilization of e-marketplaces in the logistics industry, supplemented by some practical examples.

Illés, B. and Illés, B.: *Agribusiness Clusters and Logistic Processes Through the Example of Hungary*. This paper shows a basic summary of the logistic processes that should be considered when decision-makers are to establish a hub-and-spoke type business network. Tools for creating a hub-and-spoke network for waste management can be implemented for different types of wastes as well. The reason for choosing this area is the commitment of the authors to the agricultural development of Hungary.

The book makes an effort to ensure an equilibrium between theory and practice and to show some new approaches from the theoretical modeling aspect, as well as experimental and practical points of view.

Gabriella Bognár
Tibor Tóth

Contents

Parameter Estimation for Symmetric Stable Distributions by Probability Integral Transformation

Csilla Csendes and Sándor Fegyverneki

Abstract In this article a new parameter estimation method for symmetric stable distributions is presented, which is a variation of maximum likelihood type M-estimators. The estimator provides a joint estimation of shape parameter α, scale parameter γ and location parameter δ. The proposed method possesses all known good robustness performance properties, and is more reliable than other known methods. The estimation procedure does not use the density or the characteristic function (chf) directly, hence is faster than the maximum likelihood or chf based methods. A simulation study is performed to compare the proposed estimator with other methods based on performance properties and assessing convergence of the estimators.

1 Introduction

Although the theoretical introduction of the stable distribution family goes back to the works by Levy [14] in 1925, analysts had to wait for easily applicable smooth usage in practical problems until the last decades of the century. Phenomena of higher peak and heavier tails of the density function than the normal law were captured in many empirical studies, but inefficient and cumbersome algorithms, and difficult, computationally intensive procedures prevented wide-spread usage of the stable family in data analysis. Nowadays, these difficulties tend to vanish with the increased computational capacity of computer hardware and achievements in numerical methodology.

Cs. Csendes (✉)
Department of Mathematics and Informatics, Corvinus University of Budapest, Villányi út 29–33, Budapest1118, Hungary
e-mail: csilla.csendes@uni-corvinus.hu

S. Fegyverneki
Department of Applied Mathematics, University of Miskolc, Miskolc-Egyetemvàros H-3515, Miskolc, Hungary
e-mail: matfs@uni-miskolc.hu

G. Bognár and T. Tóth (eds.), *Applied Information Science, Engineering and Technology*, Topics in Intelligent Engineering and Informatics 7, DOI: 10.1007/978-3-319-01919-2_1,

The theoretical importance of stable laws is inevitable since the distribution family provides the only possible solution to the generalization of the central limit theorem (CLT). With the assumption of common finite variance for the sum of independent, identically distributed (iid) random variables the limiting distribution is the normal law. The generalization of CLT called the domain of attraction problem arises by summing iid variables with infinite variance.

In data analysis heavier tails are usually realized when a great number of observations are aggregated with very high or nearly infinite variance. It seems reasonable to assume a relation between these characteristics of data sets and the generalized CLT. If such data are modelled application of stable laws should be considered. Data sets of this sort are collected in studies in the field of financial mathematics, e.g. price changes in high frequency trading, signal processing, or measuring the data transfer of Internet traffic.

An additional reason for using the stable distribution family is the stability property, from which the name of the family originates. Regardless of some location and scaling constants, the distribution of the sum of stable random variables results again in a random variable with the same distribution. One of the definitions is based on this property:

Definition 1 *(Broad sense) Let $X, X_1, X_2, \ldots$ be iid random variables. The distribution of X is stable, if it is not concentrated at one point and if for each n there exist constants $a_n > 0$ and b_n such that*

$$\frac{X_1 + X_2 + \ldots + X_n}{a_n} - b_n \tag{1}$$

has the same distribution as X.

Computational issues are derived mainly from the lack of known exact formula for the probability density function (pdf) or the cumulative distribution function (cdf) hence the commonly used estimation methods such as the maximum likelihood estimation cannot be directly used.

The pdf and cdf of a general stable variable is available in so-called integral representation. Numerical integration has a very high computational demand and the convergence of the formula is very slow, it is possible that around a thousand terms are necessary. Nolan presented an algorithm for the evaluation of the pdf based on Fast Fourier Transform and other diagnostic tools for distribution fitting. Proposed algorithms are collected in the STABLE program and are available on Nolan's webpage [18].

A general stable distribution can be characterized by four parameters of the characteristic function (chf). The shape parameter (characteristic exponent, index of stability) $\alpha \in [0, 2]$ describes peakedness around the mean and heaviness of the tails, skewness $\beta \in [-1, 1]$ is the symmetry parameter, scale $\gamma > 0$ is a measure of dispersion and location $\delta \in \mathbb{R}$ is the mean (if it exists).

The characteristic function of stable random variable Z is

$$\phi(u|\alpha,\beta,\gamma,\delta) = \mathrm{E}\exp(iuZ) = \exp(-\gamma^{\alpha}[|u|^{\alpha} + i\beta\eta(u,\alpha)] + iu\delta), \quad (2)$$

$$\eta(u,\alpha) = \begin{cases} -(\mathrm{sign}(u))\tan(\pi\alpha/2)|u|^{\alpha}, & \text{if } \alpha \neq 1, \\ (2/\pi)u\ln|u|, & \text{if } \alpha = 1. \end{cases}$$

Statistical procedures are often based on the characteristic function, although in general it is a complex valued function and is inconvenient to use. If $\beta = 0$ the distribution is symmetric, and the characteristic function has a much simpler form. Two special symmetric members of the family are the normal ($\alpha = 2$) and the Cauchy distribution ($\alpha = 1$), for which pdf is of course available. An additional difficulty is the non-existing variance, and with $p \geq \alpha$ even the pth moment $E[|X|^p]$ is infinite.

Basic works dealing with the topic are Uchaikin and Zolotarev [23] and Zolotarev [25] on general characteristics, Adler et al. [1], Samorodnitsky and Taqqu [21], Rachev and Mittnik [20] on applications, and Nolan [16, 17] on simulation and statistical diagnostics.

These properties of the stable family require extraordinary procedures for parameter estimation. A group of proposed methods are based on the chf such as the regression method by Koutrovelis [13] and improvement of his method by Kogon and Williams [12] and the transformations of the chf by Press [19].

Another class of estimators uses the asymptotic Pareto tail behaviour. A recent paper by Szeidl [22] gave a new tail index estimator in the general case with the help of empirical power processes. We also mention a thorough survey paper on tail index estimators by Csörgő and Viharos [3] where tail index estimators are categorized to class of universally asymptotically normal weighted doubly logarithmic least-squares estimators and a class of kernel estimators, the latter contains the Hill estimator. The Hill estimator [9] and its many modifications are easily applicable, hence they are quite well-known, but may result in objectionable α values of magnitude around 4, in contrast to the valid parameter space $\alpha \in (0, 2]$, that's why they are completely unreliable.

One paper by Nolan [16] proposes a fast numerical ML estimation, which is accurate and has good estimation properties, but is still quite slow. Papers with historical importance should also be mentioned: primarily Fama and Roll [5] and McCulloch [15]. These methods are based on empirical percentiles of the sample and cannot be recommended for real-life applications. A simulation survey on the known methods of parameter estimation was performed by Weron [24] and Borak et al. [2]. Very recent papers like Garcia et al. [7] use Bayesian inference.

In Sect. 2 we propose a new parameter estimation method for symmetric stable variables ($\beta = 0$) based on the concept of M-estimators presented originally by Huber [10]. The procedure estimates simultaneously the three unknown parameters α, γ, δ. Estimation of the shape parameter is done with the help of a core iteration estimation procedure for the location and the scale parameter. The computation of the estimators is fastened with pre-calculated rational fraction approximations with very high accuracy (i.e. 10^{-10}). In Sect. 3 numerical results for the approximated functions are presented. Section 4 is devoted to a comparative simulation study where performance of some known parameter estimation methods and our method are described. We briefly summarize our presented work in Sect. 5.

2 Probability Integral Transformation for Symmetric Stable Distributions

Maximum likelihood type estimators (M-estimators) are defined through smooth weighting functions. The simultaneous version of M-estimators (T_n, S_n) for any location-scale family is the solution of the system

$$\sum \psi\Big(\frac{x_i - T_n}{S_n}\Big) = 0, \tag{3}$$

$$\sum \chi\Big(\frac{x_i - T_n}{S_n}\Big) = 0, \tag{4}$$

where T_n and S_n are the current estimators of location parameter T and scale parameter S, x_i's are the iid. sample elements, and functions ψ and χ possess some properties that guarantee uniqueness and existence of the solution. Davies [4] gives the proper conditions for ψ and χ, under which the simultaneous estimation can be used.

We use Probability Integral Transformation (PT) and the method of moments to define our ψ and χ functions. Our notations follow the usual notations of robust statistics used by Huber [10, 11], Hampel et al. [8]. Let us consider cdf F, and F_0 of the same type, where $F(x) = F_0((x - T)/S)$, T and S is a location and a scale parameter, respectively and they are defined according to F_0. The inverse function theorem states that if cdf F of a continuous random variable ξ can be inverted, then $F(\xi)$ is uniformly distributed on $[0, 1]$. Using the inverse function method to $F_0((x - T)/S)$ we construct a new uniform variable. For the new variable the expectation and variance is

$$E_F\Big(F_0\Big(\frac{\xi - T}{S}\Big)\Big) = \frac{1}{2}, \tag{5}$$

$$D_F^2\Big(F_0\Big(\frac{\xi - T}{S}\Big)\Big) = \frac{1}{12}. \tag{6}$$

Assuming that F_0 is symmetric, $D^2(X) = E^2(X)$, the probability integral transformation estimators of T and S are the solutions of the system

$$\sum_{i=1}^{n} \Big(F_0\Big(\frac{x_i - T}{S}\Big) - \frac{1}{2}\Big) = 0, \tag{7}$$

$$\sum_{i=1}^{n}\left(\left(F_0\left(\frac{x_i - T}{S}\right) - \frac{1}{2}\right)^2 - \frac{1}{12}\right) = 0, \tag{8}$$

where x_i's are the sample elements and the equations follow from the law of large numbers.

Hence, functions ψ and χ of the simultaneous M-estimator are $\psi(x) = F_0(x) - 0.5$ and $\chi(x) = \psi^2(x) - 1/12$. If the random variable ξ has the same type of distribution as the type of F_0, then the variance is $1/12$ as in Eq. (6). If the types are not the same, the variance is

$$\begin{aligned}\mathcal{B} = D^2_{F_\xi}(\psi(\xi)) = D^2_{F_\xi}\left(F_0(\xi) - 0.5\right) \\ = \int_{-\infty}^{\infty}(F_0(x) - 0.5)^2 dF_\xi(x) = \int_{-\infty}^{\infty}(F_0(x) - 0.5)^2 f_\xi(x)dx,\end{aligned} \tag{9}$$

where f_ξ and F_ξ denote the pdf and cdf of the sample, respectively.

Then the final equation system to solve for the (T_n, S_n) location and scale estimators of (T, S) is

$$\sum \psi\left(\frac{x_i - T_n}{S_n}\right) = 0, \tag{10}$$

$$\sum \psi^2\left(\frac{x_i - T_n}{S_n}\right) = (n-1)\mathcal{B}, \tag{11}$$

where x_i, ψ are as above.

An iterative algorithm called the *ping—pong* method was constructed to solve such a system, see Fegyverneki [6], Huber [10] and Hampel et al. [8]. We present the method first in its original form, where it is assumed that the pdf of the sample to calculate $\mathcal{B}$ is completely known. For example in the case of an underlying normal distribution, the type of cdf is obviously known. The iteration is based on the modified Newton method, and its convergence follows from Banach's fixpoint theorem, because it is a contraction with probability 1. In each step Eqs. (12) and (13) are evaluated in turn until reaching an expected predefined precision.

The step for the location parameter:

$$T_n^{(m+1)} = T_n^{(m)} + \frac{1}{n} S_n^{(m)} \sum_{i=1}^{n} \psi\left(\frac{x_i - T_n^{(m)}}{S_n^{(m)}}\right), \tag{12}$$

The step for the scale parameter:

$$[S_n^{(m+1)}]^2 = \frac{1}{(n-1)\mathcal{B}} \sum_{i=1}^{n} \psi^2\left(\frac{x_i - T_n^{(m+1)}}{S_n^{(m)}}\right)[S_n^{(m)}]^2, \tag{13}$$

where the initial approximations are

$$T_n^{(0)} = med\{x_i\}, \tag{14}$$

$$S_n^{(0)} = C \cdot MAD, \tag{15}$$

where $med\{x_i\}$ is the median, MAD denotes median absolute deviation $MAD = med\{|x_i - med\{x_i\}|\}$, and $S_n^{(m)}$ and $T_n^{(m)}$ are current estimations of S and T in step m of the iteration, respectively. Constant $C = F_0^{-1}(3/4)$ is used to have the initial estimation unbiased (F_0 is symmetric). The initial approximations median and median absolute deviation are consistent robust estimates of the location and scale parameters, if constant C is properly chosen, see Huber [10].

For a symmetric stable random variable $\xi_\alpha \sim S_\alpha(0, \gamma, \delta)$ the parameter estimation procedure with the ping-pong algorithm would need the special representative of the type $F_{0,\alpha}$. Although no closed formula is known for the cdf of a general α-stable variable, we can use cdf's of two symmetric members of the stable family, Cauchy and normal distribution.

Definition 2 *(Normal distribution, $\alpha = 2$)*

$$f(\gamma, \delta; x) = \frac{1}{\sqrt{4\pi\gamma^2}} \exp\left(-\frac{(x-\delta)^2}{4\,\gamma^2}\right), \qquad \gamma > 0, \qquad \delta \in \mathbb{R}$$

Definition 3 *(Cauchy distribution, $\alpha = 1$)*

$$f(\gamma, \delta; x) = \frac{1}{\pi} \frac{\gamma}{\gamma^2 + (x-\delta)^2}, \qquad \gamma > 0, \qquad \delta \in \mathbb{R}.$$

The distribution functions are $\Phi(x)$ and $F_{Cauchy} = 1/\pi \arctan x + 1/2$. We substitute the two cdf's into $\psi(x) = F_0(x) - 0.5$, respectively. Hence, we get two different functions $\mathcal{B}$, one where F_0 is considered as the Cauchy distribution function

$$\int_{-\infty}^{\infty} \left(\frac{1}{\pi} \arctan x\right)^2 dF_\alpha = \int_{-\infty}^{\infty} \left(\frac{1}{\pi} \arctan x\right)^2 f_\alpha(x) dx = \mathcal{B}_1(\alpha), \tag{16}$$

and one where F_0 is considered as the normal distribution function

$$\int_{-\infty}^{\infty} \left(\Phi(x) - \frac{1}{2}\right)^2 dF_\alpha = \int_{-\infty}^{\infty} \left(\Phi(x) - \frac{1}{2}\right)^2 f_\alpha(x) dx = \mathcal{B}_2(\alpha), \tag{17}$$

where F_α and f_α denote the cdf and pdf of the α-stable sample.

The functions $\mathcal{B}$ involve the unknown α-stable pdf f_α in the integrand. Instead of time consuming numerical integration we use pre-calculated approximations of these functions; for a description of the calculations see Sect. 3. The functions $\mathcal{B}$ also involve the shape parameter of the sample, which is to be estimated. The functions $\mathcal{B}_1$ and $\mathcal{B}_2$ depend on the sample's α through f_α. Hence, it seems the procedure needs a guess to be made on α to calculate scale and location.

Reversing this concept, our presented algorithm estimates the shape parameter α with the help of the scale parameter approximations. Denote the scale estimators calculated with $\mathcal{B}_1(\alpha)$ and $\mathcal{B}_2(\alpha)$ in Eq. (13) by $S_1(\alpha)$ and $S_2(\alpha)$, respectively. The scale estimators can be calculated numerically with arbitrary α on the interval $[1, 2]$. Since not the proper $F_{0,\alpha}$ was used for the transformation at $\mathcal{B}$, the scale approximations $S_1(a)$ and $S_2(a)$ deviate for every $a \in [1, 2]$ except at the correct shape parameter of the sample, denoted by $\hat{\alpha}$. At the point of estimation $\hat{\alpha}$ the scale approximations according to the Cauchy and normal distribution are balanced because of the appropriate value of α, i.e. $S_1(\hat{\alpha}) = S_2(\hat{\alpha})$.

For finding that unique point we construct an iteration whose logic is similar to the bisection search, called the cut-and-try method. Initially we choose $\alpha = 1$ and $\alpha = 2$, the two valid endpoints of the parameter set for α. We calculate both $S_1(\alpha)$ and $S_2(\alpha)$ which yields four scale estimations. At the endpoints the differences of the corresponding scale estimators are greater than for α's within the interval $[1, 2]$ on the same side and the difference has an opposite sign. As one gets closer to $\hat{\alpha}$ from both sides, the deviation monotonically tends to zero.

If we consider scale approximations as functions of α then the two scale functions form two monotonically increasing curves on interval $[1, 2]$ with one intersection point which gives $\hat{\alpha}$. After initialization, we calculate the midpoint of the interval and calculate S_1 and S_2 at this point. We determine the relation of scale approximations and modify the interval for the next iteration so that the intersection point is still inside. We consider a given precision ϵ, so if the deviation $d = |a_{i-1} - a_i|$ is smaller than ϵ, we stop the algorithm. Scale estimator $\hat{\gamma}$ is the average of S_1 and S_2, location estimator $\hat{\delta}$ is the average of T_1 and T_2 at the last iteration.

Algorithm

1. Set ϵ accuracy.
2. Set $a_0 = a_L = 1$ and $a_1 = a_U = 2$ and calculate scale estimations $S_1(a_L)$, $S_2(a_L)$, $S_1(a_U)$, $S_2(a_U)$.
3. The initial conditions $S_1(a_L) < S_2(a_L)$ and $S_2(a_U) < S_1(a_U)$ are basically true, otherwise there is no intersection point and the method fails (return -1).
4. While $|a_{i-1} - a_i| > \epsilon$

   Set $a_i = (a_U + a_L)/2$ and calculate $S_1(a_i)$ and $S_2(a_i)$.
   If $S_1(a_i) < S_2(a_i)$, then set $a_L = a_i$, else set $a_U = a_i$.

5. Return a_i as $\hat{\alpha}$
6. Return $(S_1(a_i) + S_2(a_i)/2)$ as $\hat{\gamma}$, $(T_{1,i} + T_{2,i})/2$ as $\hat{\delta}$

In Figs. 1 and 2 scale parameter curves are plotted for different sample sizes and parameters α and γ. The intersection point of the curves show on the horizontal axis the estimation of α, on the vertical axis scale parameter γ. In Fig. 1 scale parameter $\gamma = 1$, sample size and parameter α are from top to bottom: 1/a. 50 elements, $\alpha = 1.2$; 1/b. 50 elements, $\alpha = 1.8$; 1/c. 500 elements, $\alpha = 1.3$; 1/d. 500 elements, $\alpha = 1.7$. In Fig. 2 parameter γ is also varying, parameters and sample size from top

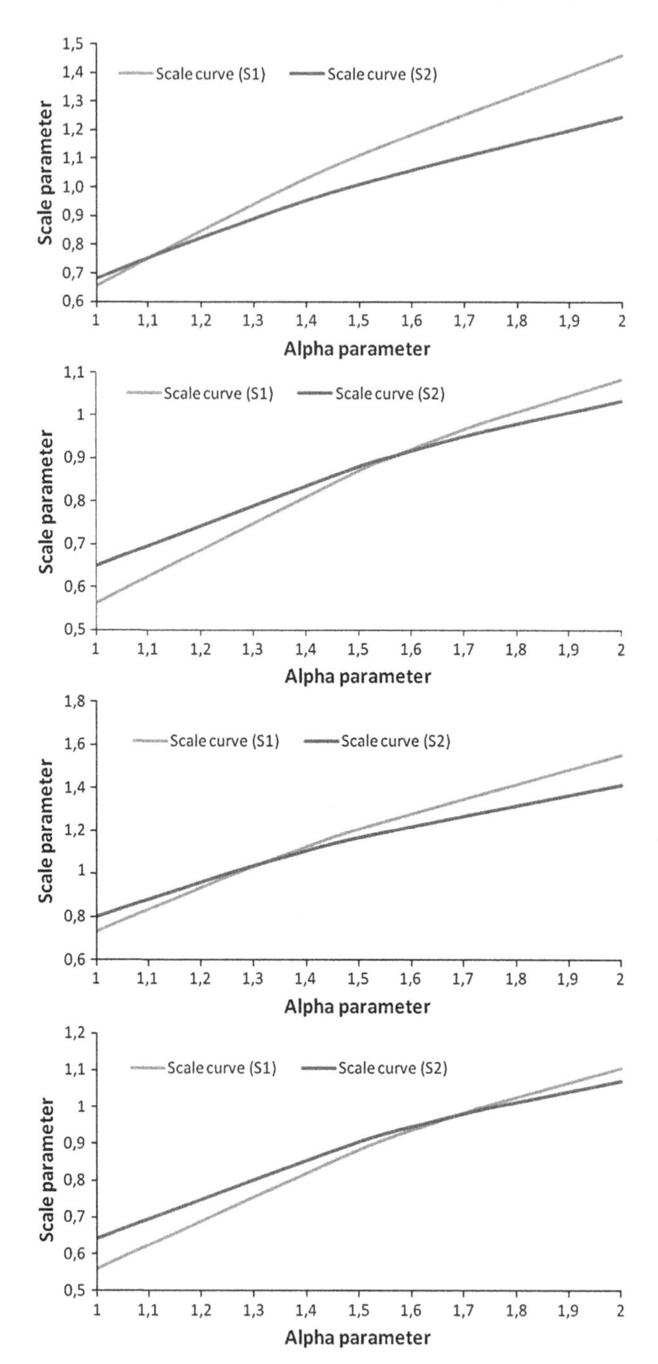

Fig. 1 $S_1(\alpha)$ and $S_2(\alpha)$ scale parameter curves computed with ping-pong method with functions $\mathcal{B}_1(\alpha)$ and $\mathcal{B}_2(\alpha)$ (1/a. sample size: 50 elements, $\alpha = 1.2$; 1/b. sample size: 50 elements, $\alpha = 1.8$; 1/c. sample size: 500 elements, $\alpha = 1.3$; 1/d. sample size: 500 elements, $\alpha = 1.7$)

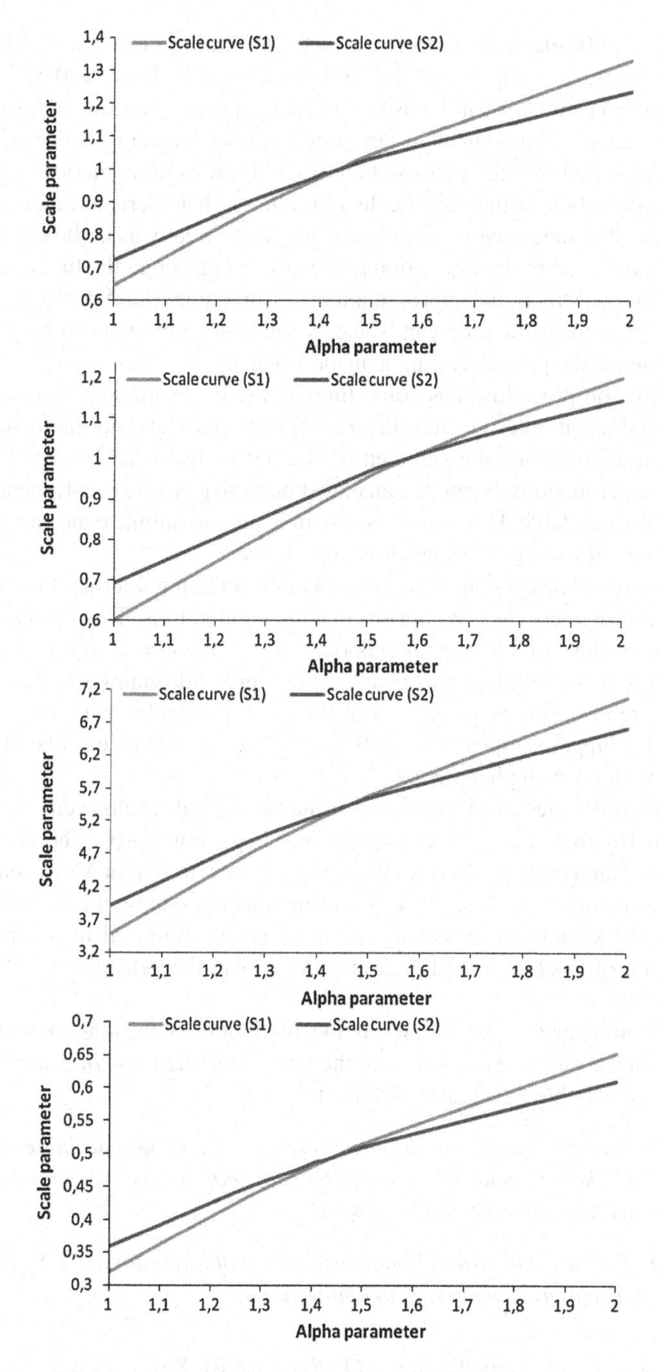

Fig. 2 $S_1(\alpha)$ and $S_2(\alpha)$ scale parameter curves computed with ping-pong method with functions $\mathcal{B}_1(\alpha)$ and $\mathcal{B}_2(\alpha)$ (2/a. sample size: 5000 elements, $\alpha = 1.4, \gamma = 1$; 2/b. sample size: 5000 elements, $\alpha = 1.6, \gamma = 1$; 2/c. sample size: 500 elements, $\alpha = 1.5, \gamma = 5$; 2/d. sample size: 500, $\alpha = 1.5$, $\gamma = 0.5$)

to bottom: 2/a. 5000 elements, $\alpha = 1.4$, $\gamma = 1$; 2/b. 5000 elements, $\alpha = 1.6$, $\gamma = 1$; 2/c. 500 elements, $\alpha = 1.5$, $\gamma = 5$; 2/d. 500 elements, $\alpha = 1.5$, $\gamma = 0.5$.

The proposed estimator can be used for real data sets of various fields of applications. It is easy to implement in any programming language, which is a great advantage compared to other proposed methods. There is no numerical integration and no approximation of the density, distribution, or characteristic function during the procedure. The unknown α-stable pdf is not used directly as in the classical ML method, it is involved in the estimation procedure only through the functions $\mathcal{B}$. We present the rational fractional approximations for functions $\mathcal{B}$ in Sect. 3 for practical use. Accuracy of the estimation can be easily set, as the iteration can be stopped at arbitrary ϵ, hence the procedure can be made faster.

The estimation procedure has some limitations. PT estimation works only for symmetric stable variables because the F_0 types (Cauchy and normal distribution) used in the transformation are symmetric. The parameter space for α is limited to $(1, 2)$, however functions $\mathcal{B}$ can be calculated numerically for $\alpha < 1$, meaning that the method is extendable. However, this case indicates an infinite expectation, which from the practical point of view has less importance.

Another issue is that for small samples the intersection point may fall outside of the interval, so in a few cases it happens that the method fails. This is more likely if parameter α is close to the interval endpoints $\alpha = 1$ or $\alpha = 2$. In our simulations approximately 6 % of replications resulted in failure with samples of 50 elements, and 1 % of replications were failures in the case of samples with 100 elements. Estimators for bigger samples ($n = 400$ or $n = 2500$) could be calculated in every replication with no such phenomena.

About the properties of M-estimators some theoretical results were proposed in Huber [10], Hampel et al. [8], Davies [4] and Fegyverneki [6]. The comprehensive book by Hampel et al. [8] describes not only how to obtain M-estimators, but also explain the different concepts of robustness and goodness measures of robust estimators. We summarize here only the main results achieved in relation to the PT estimation of location T and scale parameter S in the general case. We follow Fegyverneki [6].

Let us assume that $\xi = S\eta + T$, where the distribution of the random variable η is $G_0(x)$. Given the sample $\xi_1, \xi_2, \cdots$ and the type of distribution G_0, the distribution of the random variable ξ_i is $G_0((x - T)/S)$.

Theorem 1 *Assume that G_0 is differentiable, strictly monotone increasing, and $G_0(0) = 0.5$. Then T_n and S_n are well defined; that is Eqs.* (10) *and* (11) *equation system has a unique solution with S_n>0.*

Theorem 2 *The two-dimensional joint distribution of estimators (T_n, S_n) under the conditions of Theorem 1, converges to a normal one*

$$\sqrt{n}((T_n, S_n) - (T, S)) \xrightarrow{d} N(0, \Sigma),$$

where the covariance matrix Σ is given by $\Sigma = C^{-1}D[C^{-1}]^T$.

The matrices C and D are given by

$$C = \begin{pmatrix} E\left(\frac{\partial}{\partial T}\psi\left(\frac{\xi-T}{S}\right)\right) & E\left(\frac{\partial}{\partial S}\psi\left(\frac{\xi-T}{S}\right)\right) \\ E\left(\frac{\partial}{\partial T}\chi\left(\frac{\xi-T}{S}\right)\right) & E\left(\frac{\partial}{\partial S}\chi\left(\frac{\xi-T}{S}\right)\right) \end{pmatrix},$$

and

$$D = \begin{pmatrix} E(\psi^2(\eta)) & E(\psi(\eta)\chi(\eta)) \\ E(\psi(\eta)\chi(\eta)) & E(\chi^2(\eta)) \end{pmatrix} = \begin{pmatrix} \frac{1}{12} & 0 \\ 0 & \frac{1}{180} \end{pmatrix},$$

where $\eta \sim G_0$. The covariance matrix depends on the type of F_0.

Theorem 3 *The PT-estimators, under the conditions of Theorem 1, are B-robust, V-robust, qualitatively robust and their breakdown points are*

$$\varepsilon^*(T_n) = \frac{\delta}{1+\delta} = 0.5, \quad \textit{where} \quad \delta = \min\left\{-\frac{\psi(-\infty)}{\psi(+\infty)}, -\frac{\psi(+\infty)}{\psi(-\infty)}\right\}$$

and

$$\varepsilon^*(S_n) = \frac{-\chi(0)}{\chi(-\infty) - \chi(0)} = \frac{1}{3}.$$

For PT estimators ($\hat{\alpha}$, $\hat{\gamma}$, $\hat{\delta}$) we examined the joint normality with Kolmogorov-Smirnov and χ^2 goodness-of-fit tests but no theoretical results have been achieved yet. Our simulations showed that asymptotically the estimator of shape parameter α follows a normal distribution. For small samples ($n = 50$, $n = 100$), with Monte-Carlo replications $r = 2500$ the normality does not hold. We examined performance properties of the estimators and correlation around 0.4–0.6 between α and γ was revealed.

3 Numerical Approximation of Functions $\mathcal{B}$

In this section we give a detailed description of how the functions $\mathcal{B}_1$ and $\mathcal{B}_2$ was approximated and present the coefficients necessary to use our procedure in practice. Recall that function $\mathcal{B}$ is an expectation and from the law of large numbers can be approximated with the average:

$$\mathcal{B}_1(\alpha) \approx \frac{1}{n}\sum_{i=1}^{n}(\Phi(x_i) - 0.5)^2, \tag{18}$$

$$\mathcal{B}_2(\alpha) \approx \frac{1}{n}\sum_{i=1}^{n}\left(\frac{1}{\pi}\arctan x_i\right)^2, \tag{19}$$

Table 1 Approximated values of $\mathcal{B}_1$ and $\mathcal{B}_2$ depend on α

α	$\mathcal{B}_1(\alpha)$	$\mathcal{B}_2(\alpha)$
1	0.0833333333333333	0.126807877965645
1.1	0.0758844534723818	0.118966259082521
1.2	0.0697612892957584	0.112284032323310
1.3	0.0646999988570841	0.106570898511029
1.4	0.0604648399039825	0.101682622480835
1.5	0.0569093151515006	0.097443890906153
1.6	0.0538933607717261	0.093798682214659
1.7	0.0513226066667932	0.090637100836610
1.8	0.0491126022363082	0.087875629036068
1.9	0.0472087085432832	0.085445768785679
2	0.0455654051822800	0.083333333333333

where x_i denotes the α-stable sample elements and corresponds to the arguments of $\mathcal{B}_i$,
$i = 1, 2$. These averages were calculated by α-stable random samples with 20 million elements and α ranging with step 0.1 from 1 to 2.

A symmetric α-stable standard ($\gamma = 1$, $\delta = 0$) variable Z is generated as

$$Z(\alpha, 0, 1, 0) = \frac{sin(\alpha\xi)}{(cos\xi)^{\frac{1}{\alpha}}}\left(\frac{cos((1-\alpha)\xi)}{\eta}\right)^{\frac{1-\alpha}{\alpha}}, \tag{20}$$

where η is a standard exponential variable, and ξ is uniform on $(-\pi/2, \pi/2)$. For standardized variables $Z(\alpha, 0)/\alpha^{\frac{1}{\alpha}}$ can be used. This method is due to Zolotarev [25].

In Table 1 the approximations according to Eqs. (18) and (19) are shown. The very high number of generated random elements guarantees precision. The functions $\mathcal{B}_i$ were approximated afterwards with a rational fraction function in the form

$$\mathcal{B}_i(x) = \frac{a_5x^5 + a_4x^4 + a_3x^3 + a_2x^2 + a_1x + a_0}{x^4 + b_3x^3 + b_2x^2 + b_1x + b_0}, \tag{21}$$

where $i = 1, 2$.

The approximated values of the functions $\mathcal{B}_i$ at the base points $\alpha = 1, 1.1, \ldots, 2$ are substituted into the left side and the base points are substituted into the right side of Eq. (21). After arrangement we have eleven linear equations, where the coefficients are the unknowns. The linear equation system for the coefficients is overdetermined. We found the solution of the equation system decreased by one equation, for which the least squared differences from the values at base points were the smallest. Table 2 shows the coefficients a_5, ..., a_0, b_3, ..., b_0 for the rational fraction approximation.

Table 2 Coefficients for the rational fraction function

Coefficient	$\mathcal{B}_1$	$\mathcal{B}_2$
b_3	−3.83008202167381	−5.44424585925350
b_2	4.78393407388667	8.41128641608921
b_1	−2.07519730244991	−0.91519048820337
b_0	0.18011293964047	−4.11676503125739
a_5	0.00557315701358	0.02536047583564
a_4	−0.02655295929925	−0.20581738159343
a_3	0.12169846973572	0.84848196461615
a_2	−0.31598423581221	−2.18605774383017
a_1	0.34060543137722	3.06692780009580
a_0	−0.12044269481921	−1.68393473522529

4 Performance and Comparison

In this section we investigate performance properties of the proposed PT estimation procedure by simulations. Computational times and a table of confidence intervals for the estimator $\hat{\alpha}$ based on a simulation sequence with various α parameters and sample sizes are presented. We also present a simulation sequence very similar to the simulations by Weron [24] in order to compare our procedure to parameter estimation methods investigated in that paper.

Computational times. We present tables about run times of the PT estimation procedure. The program was implemented in Java programming language and the properties of the computer are: Intel Core(R) Quad(TM) Q9400, 2.66 GHz CPU, 4 GB RAM, Windows 7 64 bit operation system. We simulated samples of $n = 50, 100, 400, 2500$ elements with $\alpha = 1.3, 1.5, 1.7$. The precision of the ping-pong method is $\epsilon = 10^{-7}$ and the number of iterations in the cut-and-try method is 18, which yields a precision of 2^{-18}. The running times include only the time spent on parameter estimation.

In Table 3 the running times of Monte-Carlo replications $r = 100$ are summarized in minutes by parameter α and sample size n. The parameter estimation procedure has shorter run times for samples with higher α according to the simulation. The reason is that the sample elements are spread on a smaller interval for higher α and the convergence of the scale estimator is faster. Furthermore, the run time of the algorithm seems to have a linear relation to the sample size as it is seen in Fig. 3.

Confidence interval for estimator $\hat{\alpha}$. Table 4 shows confidence intervals for $\hat{\alpha} - \alpha$ and the length of the intervals for significance levels 90, 95, 99 %. The confidence intervals are calculated with $r = 2500$ Monte-Carlo replications according to the formula of the t-statistic.

Comparison. According to our simulation results PT estimation coincides in performance with the methods investigated in Weron [24], namely Fama and Roll's procedure (FR), McCulloch's (CULL) and regression type (REG) estimators, and the method of moments (MOM). Tables of Ref. [24, Tables 4.5 and 4.6] contain

Table 3 Running times of PT algorithm with different α and sample size n in minutes, Monte-Carlo replication $r = 100$

Sample size (n)	$\alpha = 1.3$	$\alpha = 1.5$	$\alpha = 1.7$
n= 50	0.69	0.65	0.53
n= 100	1.35	1.22	1.13
n= 400	4.72	4.28	4.01
n= 2500	29.20	23.79	20.82

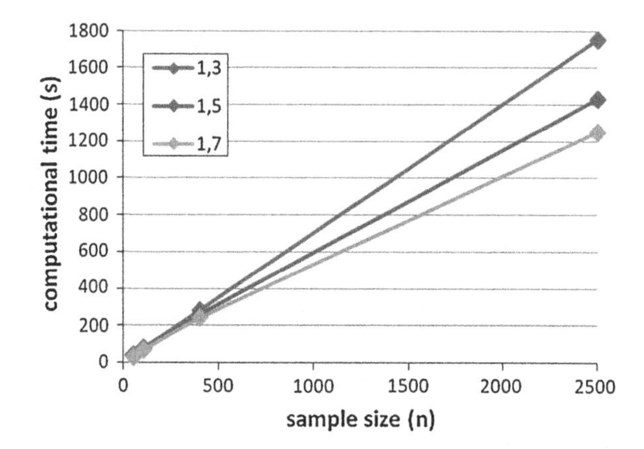

Fig. 3 Computational time of PT algorithm for different α and sample size n in seconds, Monte-Carlo replications $r = 100$

Table 4 Confidence intervals in simulation sequence, with t-statistic (CI_L and CI_U are lower and upper limits of the intervals)

α	n	$90\,\%CI_L$	$90\,\%CI_U$	CI len.	$95\,\%CI_L$	$95\,\%CI_U$	CI len.	$99\,\%CI_L$	$99\,\%CI_U$	CI len.
1.3	50	0.0556	0.0693	0.0069	0.0543	0.0706	0.0082	0.0517	0.0732	0.0108
	100	0.0237	0.0338	0.0051	0.0227	0.0348	0.0060	0.0208	0.0367	0.0079
	400	0.0042	0.0091	0.0024	0.0038	0.0095	0.0029	0.0029	0.0104	0.0038
	2500	0.0009	0.0029	0.0010	0.0007	0.0031	0.0012	0.0004	0.0035	0.0015
1.5	50	0.0282	0.0424	0.0071	0.0268	0.0437	0.0085	0.0242	0.0464	0.0111
	100	0.0192	0.0295	0.0052	0.0182	0.0305	0.0062	0.0162	0.0325	0.0081
	400	0.0059	0.0109	0.0025	0.0054	0.0114	0.0030	0.0044	0.0124	0.0040
	2500	0.0003	0.0023	0.0010	0.0001	0.0025	0.0012	−0.0003	0.0029	0.0016
1.7	50	−0.0109	0.0019	0.0064	−0.0121	0.0031	0.0076	−0.0145	0.0055	0.0100
	100	0.0054	0.0150	0.0048	0.0045	0.0159	0.0057	0.0027	0.0176	0.0075
	400	0.0028	0.0077	0.0024	0.0023	0.0081	0.0029	0.0014	0.0091	0.0038
	2500	−0.0008	0.0011	0.0010	−0.0010	0.0013	0.0012	−0.0014	0.0016	0.0015

Table 5 Comparison between estimation procedures ($\alpha = 1.5, \gamma = 1, \delta = 0, n = 500, r = 25$)

γ	Method	$\overline{\alpha}$	α_{min}	α_{max}	MSE_{α}	$\overline{\gamma}$	γ_{min}	γ_{max}	MSE_{γ}
$\gamma = 10$	FR	1.502	1.298	1.745	0.01128	10.230	8.989	11.291	0.35687
	CULL	1.502	1.288	1.771	0.01491	10.100	9.073	11.095	0.33746
	REG	1.486	1.245	1.628	0.00972	10.067	8.999	10.937	0.28761
	PT	1.512	1.382	1.682	0.00467	9.937	8.829	11.592	0.40294
$\gamma = 1$	FR	1.518	1.351	1.677	0.00558	1.020	0.908	1.205	0.00447
	CULL	1.518	1.415	1.692	0.00553	1.017	0.903	1.183	0.00394
	REG	1.515	1.332	1.643	0.00729	1.011	0.914	1.148	0.00271
	PT	1.495	1.319	1.663	0.00630	0.989	0.850	1.080	0.00296
$\gamma = 0.1$	FR	1.512	1.374	1.643	0.00595	0.099	0.088	0.110	0.00003
	CULL	1.498	1.370	1.603	0.00437	0.098	0.087	0.108	0.00003
	REG	1.495	1.361	1.620	0.00549	0.098	0.086	0.108	0.00004
	PT	1.512	1.360	1.594	0.00395	0.100	0.088	0.107	0.00001

Source of FR, CULL and REG estimation data is Ref. [24, Table 4.5]

Table 6 Comparison between estimation procedures ($\gamma = 1, \delta = 0, n = 2000, r = 25$)

α	Method	$\overline{\alpha}$	α_{min}	α_{max}	MSE_{α}	$\overline{\gamma}$	γ_{min}	γ_{max}	MSE_{γ}
$\alpha = 1.8$	FR	1.812	1.726	1.950	0.00306	0.993	0.952	1.031	0.00055
	CULL	1.817	1.714	2.000	0.00475	0.994	0.938	1.032	0.00054
	MOM	1.800	1.745	1.856	0.00092	0.992	0.940	1.029	0.00038
	REG	1.801	1.735	1.848	0.00101	0.992	0.948	1.023	0.00033
	PT	1.795	1.751	1.846	0.0008	0.999	0.956	1.049	0.0007
$\alpha = 1.5$	FR	1.496	1.410	1.588	0.00260	1.003	0.946	1.047	0.00073
	CULL	1.498	1.367	1.595	0.00288	1.001	0.931	1.051	0.00104
	MOM	1.487	1.409	1.561	0.00208	0.999	0.953	1.048	0.00061
	REG	1.492	1.371	1.630	0.00351	1.002	0.948	1.053	0.00082
	PT	1.504	1.454	1.574	0.0012	1.004	0.952	1.060	0.0007
$\alpha = 1.2$	FR	1.207	1.159	1.264	0.00099	0.994	0.935	1.060	0.00082
	CULL	1.211	1.135	1.275	0.00143	0.985	0.945	1.038	0.00096
	MOM	1.204	1.111	1.276	0.00148	0.993	0.909	1.044	0.00095
	REG	1.201	1.132	1.272	0.00122	0.992	0.927	1.071	0.00099
	PT	1.195	1.116	1.264	0.0017	0.997	0.915	1.068	0.0017

Source of FR, CULL, MOM and REG estimation data is Ref. [24, Table 4.6]

simulation results for different α, β with $n = 100$, $n = 500$, $n = 2000$ and Monte-Carlo replication $r = 25$, $r = 50$. We extracted the simulation results for the symmetric cases from [24] and ran PT estimation with those simulation properties. We compared the average, minimum, maximum, and MSE values of the PT estimators $\hat{\alpha}$ and $\hat{\gamma}$ to the other estimators presented in Tables 5, 6, 7 and 8.

In Table 5 the scale parameter γ varies as $\gamma = 0.1, 1, 10$. In Tables 6, 7 and 8 simulation results are shown where standard samples ($\gamma = 1$, $\delta = 0$) were generated and in each case $\alpha = 1.2, 1.5, 1.8$ was investigated with different sample sizes and

Table 7 Comparison between estimation procedures ($\alpha = 1.5, \gamma = 1, \delta = 0, n = 500, r = 25$)

α	Method	$\overline{\alpha}$	α_{min}	α_{max}	MSE_α	$\overline{\gamma}$	γ_{min}	γ_{max}	MSE_γ
$\alpha = 1.8$	FR	1.794	1.597	1.989	0.00978	1.007	0.912	1.124	0.00316
	CULL	1.806	1.597	2.000	0.01605	1.011	0.891	1.131	0.00378
	MOM	1.776	1.560	1.909	0.00630	1.003	0.897	1.110	0.00243
	REG	1.774	1.628	1.887	0.00528	1.003	0.895	1.108	0.00250
	PT	1.818	1.716	1.911	0.0027	1.011	0.894	1.084	0.0019
$\alpha = 1.5$	FR	1.497	1.329	1.733	0.00833	0.992	0.900	1.109	0.00285
	CULL	1.487	1.274	1.690	0.01077	0.983	0.883	1.091	0.00320
	MOM	1.501	1.277	1.698	0.00677	0.996	0.847	1.110	0.00389
	REG	1.504	1.338	1.663	0.00664	0.991	0.894	1.147	0.00361
	PT	1.498	1.346	1.610	0.0046	1.010	0.908	1.156	0.0030
$\alpha = 1.2$	FR	1.174	1.063	1.353	0.00379	0.995	0.867	1.100	0.00442
	CULL	1.169	1.041	1.329	0.00470	0.982	0.883	1.078	0.00420
	MOM	1.155	1.017	1.299	0.00692	0.970	0.857	1.115	0.00459
	REG	1.195	1.059	1.347	0.00614	1.000	0.870	1.143	0.00486
	PT	1.218	1.092	1.332	0.0043	0.996	0.865	1.063	0.0027

Source of FR, CULL, MOM and REG estimation data is Ref. [24, Table 4.6]

Table 8 Comparison between estimation procedures ($\alpha = 1.5, \gamma = 1, \delta = 0, n = 100, r = 50$)

α	Method	$\overline{\alpha}$	α_{min}	α_{max}	MSE_α	$\overline{\gamma}$	γ_{min}	γ_{max}	MSE_γ
$\alpha = 1.8$	FR	1.779	1.217	2.000	0.03902	0.987	0.803	1.233	0.01037
	CULL	1.788	1.284	2.000	0.04584	0.988	0.742	1.168	0.00957
	MOM	1.828	1.427	2.000	0.02498	1.001	0.840	1.185	0.00756
	REG	1.812	1.448	2.000	0.02528	0.995	0.824	1.182	0.00852
	PT	1.779	1.488	1.999	0.0191	0.996	0.788	1.135	0.0077
$\alpha = 1.5$	FR	1.497	1.140	2.000	0.04066	1.004	0.679	1.374	0.02051
	CULL	1.504	1.113	2.000	0.04872	0.992	0.676	1.378	0.01980
	MOM	1.527	1.116	1.958	0.04003	1.016	0.646	1.320	0.01970
	REG	1.495	0.955	1.909	0.04424	0.998	0.654	1.313	0.01939
	PT	1.516	1.244	1.875	0.0169	1.034	0.715	1.229	0.0140
$\alpha = 1.2$	FR	1.165	0.887	1.453	0.02202	0.995	0.800	1.264	0.01196
	CULL	1.150	0.870	1.460	0.02429	0.949	0.693	1.161	0.01371
	MOM	1.198	0.848	1.564	0.02624	0.992	0.721	1.273	0.01654
	REG	1.208	0.865	1.543	0.01851	1.004	0.754	1.289	0.01255
	PT	1.237	1.023	1.620	0.0204	1.011	0.758	1.368	0.0191

Source of FR, CULL, MOM and REG estimation data is Ref. [24, Table 4.6]

replications. To summarize the results, it can be stated that the accuracy of the PT estimator does not deviate significantly from the other methods.

5 Summary

In this chapter a robust parameter estimation method is presented for symmetric stable distributions based on simultaneous M-estimators. The procedure estimates jointly the three unknown parameters α, γ, δ with an iterative algorithm. All necessary constants and function approximations are presented for numerical evaluation of the estimators. The method can be useful for practical applications, because it can be implemented easily. The PT estimators have all the known good properties of robust M-estimators. A simulation study is presented to compare the PT estimators to some known methods in the field.

Acknowledgments The described work was carried out as part of the TÁMOP-4.2.2/B-10/1-2010-0008 project in the framework of the New Hungarian Development Plan. The realization of this project is supported by the European Union, co-financed by the European Social Fund.

References

1. Adler, J.R., Feldman, R.E., Taqqu, M.S. (eds.): A Practical Guide to Heavy Tails: Statistical Techniques and Applications. Birkhäuser, Boston (1998)
2. Borak, Sz., Härdle, W., Weron, R.: Stable Distributions, SFB 649 Discussion Papers, SFB649DP2005-008, Sonderforschungsbereich 649, Humboldt University, Berlin, Germany (2005)
3. Csörgő, S., Viharos, L.: Estimating the tail index. In: Szyszkowicz, B. (ed.) Asymptotic Methods in Probability and Statistics, pp. 833–881. Elsevier Science, North-Holland (1998)
4. Davies, P.L.: On locally uniformly linearizable high breakdown location and scale functionals. Ann. Stat. **26**, 1103–1125 (1998)
5. Fama, E.F., Roll, R.: Parameter estimates of symmetric stable distributions. J. Am. Stat. Assoc. **66**, 331–338 (1971)
6. Fegyverneki, S.: Robust estimators and probability integral transformations. Math. Comput. Modelling **38**, 803–814 (2003)
7. Garcia, R., Renault, E., Veredas, D.: Estimation of stable distributions by indirect inference. J. Econometrics **161**, 325–337 (2011)
8. Hampel, F.R., Ronchetti, E.M., Rousseeuw, P.J., Stahel, W.A.: Robust Statistics—The Approach Based on Influence Functions. Wiley, New York (1986)
9. Hill, B.M.: A simple general approach to inference about the tail of a distribution. Ann. Stat. **3**, 1163–1174 (1975)
10. Huber, P.J.: Robust Statistics. Wiley, New York (1981)
11. Huber, P.J.: Robust estimation of a location parameter. Ann. Math. Stat. **35**, 73–101 (1964)
12. Kogon, S.M., Williams, D.B.: Characteristic function based estimation of stable parameters. In: Adler, R., Feldman, R., Taqqu, M. (eds.) A Practical Guide to Heavy Tails: Statistical Techniques and Applications, pp. 311–335. Birkhauser, Boston (1998)
13. Koutrouvelis, I.A.: Regression-type estimation of the parameters of stable laws. J. Am. Stat. Assoc. **75**, 918–928 (1980)
14. Levy, P.: Calcul des Probabilités. Gauthier-Villars, Paris (1925)
15. McCulloch, J.H.: Simple consistent estimators of stable distribution parameters. Commun. Stat.—Simulation **15**(4), 1109–1136 (1986)
16. Nolan, J.P.: Maximum likelihood estimation of stable parameters. In: Barndorff-Nielsen, O.E., Mikosch, T., Resnick, S.I. (eds.) Levy Processes: Theory and Applications, pp. 379–400. Birkhäuser, Boston (2001)

17. Nolan, J.P.: Numerical calculation of stable densities and distribution functions. Commun. Stat.—Stoch. Model. **13**, 759–774 (1997)
18. Nolan, J.P.: website on stable distributions: http://academic2.american.edu/jpnolan/stable/stable.html
19. Press, S.J.: Applied Multivariate Analysis. Holt, Rinehart and Winston, New York (1972)
20. Rachev, S.T., Mittnik, S.: Stable Paretian Models in Finance. Wiley, New York (2000)
21. Samorodnitsky, G., Taqqu, M.: Stable Non-Gaussian Random Processes. Chapman and Hall, New York (1994)
22. Szeidl, L.: Non-normal limit theorem for a new tail index estimation. Annales Univ. Sci. Budapest, Sect. Comp. **24**, 307–322 (2004)
23. Uchaikin, V.V., Zolotarev, V.M.: Chance and Stability—Stable Distributions and their Applications. VSP, Utrecht (1999)
24. Weron, R.: Performance of the Estimators of Stable Law Parameters, Hugo Steinhaus Center for Stochastic Methods, Research Report HSC/95/1 (1995)
25. Zolotarev, V.M.: One-dimensional Stable Distributions. Translations of Mathematical Monographs, vol 65. American Mathematical Society, Providence (1986)

Prediction of the Network Administration Course Results Based on Fuzzy Inference

Zsolt Csaba Johanyák and Szilveszter Kovács

Abstract The prediction of the number of students who will pass or fail the exams in the case of a subject can be very useful information for resource allocation planning purposes. In this chapter, we report on the development of a fuzzy model, that based on the previous performance of currently enrolled students, gives a prediction for the number of students who will fail the exams of the Network Administration course at the end of the autumn semester. These students will usually re-enroll for the course in the spring semester and, conforming to previous experience, will constitute the major part of the enrolling students. The fuzzy model uses a low number of rules and applies a fuzzy rule interpolation based technique (Least Squares based Fuzzy Rule Interpolation) for inference.

1 Introduction

The introduction of the credit system in Hungarian higher education brought several advantages to the students and the institutions as well. However, the increased flexibility and eligibility put also greater responsibility on the shoulders of the students, and as a side effect it contributed to an increase in the average time necessary for the fulfillment of the academic requirements for graduations. It also made resource allocation planning more difficult for the institutional side.

This problem led particularly to a demand for the prediction of the results in the case of the Network Administration course, which is a laboratory-intensive course of the BC program in computer science where students can learn and experience in

Z. C. Johanyák
Kecskemét College, Department of Information Technologies, Izsáki út 10, Kecskemét 6000, Hungary

S. Kovács (✉)
University of Miskolc, Department of Information Technologies, Miskolc-Egyetemváros 3515, Hungary
e-mail: szkovacs@iit.uni-miskolc.hu

G. Bognár and T. Tóth (eds.), *Applied Information Science, Engineering and Technology*, Topics in Intelligent Engineering and Informatics 7, DOI: 10.1007/978-3-319-01919-2_2,

small groups the topics and practical tricks of the configuration and administration of Windows, Linux and hybrid computer networks. In order to ensure a uniform load of the laboratories and teaching staff over two semesters we needed an early estimation of the number of students who will pass or fail the exam at the end of the autumn semester. This is because those students who fail in the autumn semester usually make up the majority of the students who will enrol for this course in the spring semester next year. The prediction is made based on some historical and actual data.

Computational intelligence comprises a wide range of methods (like fuzzy reasoning [2, 8, 18–20], neural networks [22], and evolutionary techniques [13]) that can be easily used for numerical data-based prediction tasks. In this chapter, we report the application of a fuzzy rule interpolation-based inference technique and a rule base optimization method based on the clonal selection principle that ensures the creation of a low complexity fuzzy system with a reduced number of rules.

The rest of this chapter is organized as follows. Section 2 reviews briefly the fuzzy inference technique applied. Section 3 presents the main ideas of the rule base identification method used. Section 4 reports the results of the modeling and conclusions are drawn in Sect. 5.

2 Fuzzy Inference

Fuzzy reasoning systems with multidimensional input spaces can usually ensure full coverage of the input space by rule antecedents with a relatively high number of rules. This number (N_R) depends on two factors, the resolution of the partitions (the number of fuzzy sets in a partition) and the number of input dimensions. It can be calculated by

$$N_R = n_1 \cdot n_2 \cdot \ldots n_k, \tag{1}$$

where n_i is the number of fuzzy sets in the ith input dimension and k is the number of input dimensions.

The high number of rules can increase system complexity, the memory demand of the software, and the inference time as well. Besides, in some cases, due to lack of information not all the rules that describe the exact relation between the input and output of the modeled phenomena can be identified directly by the human experts or by the algorithm which extracts them automatically from experimental data.

Fuzzy rule interpolation (FRI) based inference methods were developed for the above-mentioned cases, aiming at reasoning even from those input values for which a predefined rule does not exist. The research and development of the FRI field was initiated by Kóczy at the beginning of the 1990s (e.g. [24]) and since then several techniques have been developed (e.g. [1, 6, 7, 11, 12, 14–17, 23]). Among these techniques is the Least Squares based Fuzzy Rule Interpolation (LESFRI) method [14], which determines the fuzzy conclusion from the observation (input value) in two steps. First, it generates a new rule for the current input, i.e. the reference points of the rule's antecedent sets will be identical with the reference points of the observation sets in the corresponding dimensions. Next, it determines the conclusion firing the new rule, i.e. it applies a single rule reasoning approach. We chose to use this method for our experiments owing to its interpolation and extrapolation capabilities and its availability in the free FRI Matlow Toolbox [9], as well as owing to our good experience with its previous applications.

The method LESFRI (Least Squares based Fuzzy Rule Interpolation) [14] determines the fuzzy conclusion from the observation (input value) in two steps. First, it generates a new rule for the current input, i.e. the reference points of the rule's antecedent sets will be identical with the reference points of the observation sets in the corresponding dimensions. Next, it determines the conclusion firing the new rule, i.e. it applies a single rule reasoning approach.

In the first step the antecedent and consequent sets of the new rule are calculated in each dimension by a set interpolation technique called FEATLS (Fuzzy set interpolation Technique based on the method of Least Squares). Its basic idea is that all the sets of the partition are shifted horizontally in order to reach the coincidence between their reference points and the interpolation point (reference point of the fuzzy input/output in the current dimension)(see Figs. 1 and 2).

Next, the shape of the new linguistic term is calculated from the overlapped set shapes preserving the characteristic shape type of the partition (singleton, triangle, trapezoid, polygonal, etc.) applying the method of least squares. For example in case of each breakpoint of the left flank the sum

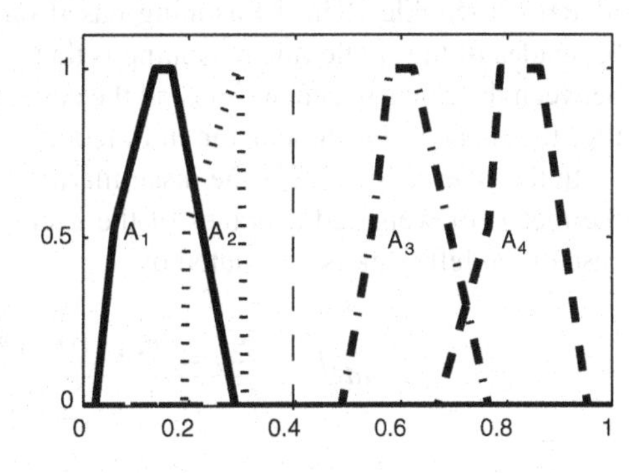

Fig. 1 Original partition and interpolation point at 0.4

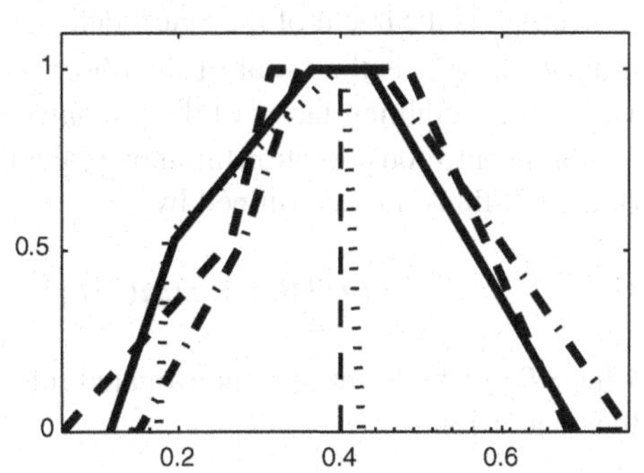

Fig. 2 Virtually shifted sets with their reference points at 0.4

$$Q_j^L = \sum_{l=1}^{n} w_l \cdot (x_{lj}^L - x_j^L)^2, \tag{2}$$

is minimized, where Q_j^L is the sum corresponding to the jth point of the left flank of the interpolated set, w_l is the weighting factor of the lth linguistic term of the partition, x_{lj}^L is the abscissa of the jth point of the left flank of the lth set and x_j^L is the abscissa of the jth point of the left flank of the interpolated set. The right flank is calculated similarly.

The position of the consequent sets is calculated independently in each output dimension by an adapted version of the crisp Shepard interpolation [21]. Thus the reference point of the conclusion is calculated as a weighted average of the reference points of the consequent sets of the known rules.

$$RP(B^i) = \frac{\sum_{l=1}^{N} RP(B_l) \cdot s_l}{\sum_{l=1}^{N} s_l}, \tag{3}$$

where $RP(B^i)$ is the reference point of the interpolated consequent set, N is the number of the rules, $RP(B_l)$ denotes the reference point of the current rule, s_l is the weight attached to the lth rule. The weighting factor is an arbitrary distance function, usually the reciprocal value of the square of the distance.

In the second step of the method the conclusion is determined by the technique SURE-LS (Single RUle REasoning based on the method of Least Squares). The basic idea of the single rule reasoning is that is measures the similarity/dissimilarity between the rule antecedent sets and the corresponding observation sets and it modifies the shape of the conclusion in function of this dissimilarity.

In the case of SURE-LS the dissimilarity on the antecedent side is calculated by the means of weighted averages of the α-cut endpoint distances (ad_α^{aL}), which in case of the left flank is calculated by

$$ad_\alpha^{aL} = \frac{\sum_{j=1}^{n_a} we_j \cdot (inf\{A_{\alpha j}^i\} - inf\{A_{\alpha j}^*\})}{\sum_{j=1}^{n_a} we_j}, \tag{4}$$

where $A_{\alpha j}^i$ is the α-cut of the antecedent set of the interpolated rule in the jth input dimension, $A_{\alpha j}^*$ is the α-cut of the observation set in the jth input dimension, and we_j is the weighting factor of the jth antecedent dimension.

Taking into consideration the above calculated differences the characteristic points of the left flank are determined by

$$inf\{B_\alpha^*\} = min(inf\{B_\alpha^i\} - ad_\alpha^{aL}, RP(B^*)), \tag{5}$$

where B_α^i is the α-cut of reinterrogated rule's consequent set and B_α^* is the α-cut of the conclusion.

3 Rule Base Identification

The performance of a fuzzy reasoning system depends strongly on the correctness of the underlying knowledge base. The rule base can be created by human experts based on their experiences, automatically from sample data by an algorithm (e.g. [3, 10, 13, 25]), or using a combination of the two options presented above (a hybrid approach).

In a hybrid approach the human experts create an initial rule base whose parameters will be modified in the course of the optimization process, which usually applies a global and/or a local search technique. Parameters of a fuzzy system can be e.g. those values which describe the fuzzy sets referred to in the antecedent and consequent parts of the rules (such as the reference points of the sets or the abscissa values of the breakpoints).

Another option for the parameter selection represents the identifiers of the language variables used in the rules. Here the task of the optimization procedure is to determine e.g. which fuzzy sets in the consequent parts of the rules ensure the best system performance.

In the course of this project we applied the hybrid approach and the above mentioned last parametrization model. By changing only the fuzzy sets referred in the rule consequents we could ensure semantically good interpretable partitions in the antecedent and consequent dimensions as well. Generally in such cases the search space of the optimal parameter set consists of

$$n_S = (N_R)^{n_o}, \tag{6}$$

discrete points, where N_R is the number of the rules, and n_o is the number of fuzzy sets in the output partition.

The tuning was done by an artificial immune system (AIS) algorithm [13]. The Clonal Selection Algorithm mimics the biological Clonal Selection Principle [4] by implementing mechanisms like clonal selection, clonal expansion, and somatic hypermutation. The first version of the algorithm was suggested by de Castro and von Zuben [5].

The algorithm generates several parameter sets as candidate solutions of the optimization problem. They are called antibodies in the AIS terminology. The initial pool of antibodies is generated as follows. One antibody is created from the initial fuzzy system and contains a description of the system's parameters. From the first instance $N - 1$ copies (clones) are made, where N is the pool size, a parameter of the method. In order to increase the diversity n_R instances are subjected to a hypermutation with a rate of p, where $p \in (0, 1)$ is an arbitrary parameter of the method. The mutation is done by changing the values of some parameters. For each parameter of an antibody a random r ($r \in (0, 1]$) value is generated. If $r < p$ a new value is selected for the parameter randomly from the pool of eligible values.

Next is the calculation of the affinity of each antibody (performance of the fuzzy system represented by the antibody), and the antibodies are sorted in descending order based on their affinity. The best ones (the first n instances) are selected for cloning. Cloning means that

$$N_c = round(\beta \cdot N) \tag{7}$$

copies are made from each selected antibody [4], where $\beta \in (0, 1]$ is a user parameter. The clones are subjected to an intensive mutation (hypermutation) which process is also called maturation of the antibodies. It performs a local search in the neighborhood of each instance belonging to the elite group. The neighborhood is closer in the case of antibodies with better performance than in the case of antibodies showing worse affinity. In this case the rate is determined by the formula

$$p = \frac{1}{e^{\rho \cdot PI}}, \tag{8}$$

where PI ($PI \in [0, 1]$) is the performance indicator of the fuzzy system and ρ ($\rho \in (0, 1]$) is a parameter of the method. In case of PI, the lowest possible value corresponds to the worst performance while the highest value indicates the best performance. The affinity of the new antibodies is also measured by means of the fuzzy system's performance.

The resulting pool of antibodies is attached to the original group and the whole repertoire is sorted in descending order based on the affinity values. The first $N - d$ antibodies (where d is parameter of the method) are selected for the next generation and the rest of them are dropped. Then d new instances are created with random bit patterns in order to ensure the diversity of the repertoire. This randomness ensures a global search character to the method.

The algorithm stops when at least one of the following conditions is met.

- n_{af} the number of affinity evaluations exceeds an upper limit—owing to the fact that the calculation of the performance indicator is one of the computationally most expensive steps,
- n_g the number of generations exceeds an upper limit,
- PI_{tr} the affinity of at least one antibody becomes greater than an upper thresh-old value,
- n_{ni} the number of consecutive generations without any improvement regarding the best affinity value (best antibody) exceeds an upper limit.

Thus the optimization method has at least eight parameters, where the first seven are: N—the number of antibodies in the repertoire (size of the repertoire), n_R—the number of antibodies selected for hypermutation in the first generation, p—the initial hyperlactation rate, n—the number of antibodies selected for cloning, β—the coefficient that determines the number of clones that are created for each selected antibody, ρ—the coefficient which determines to what extent the affinity of an antibody influences the probability of mutation, and d—the number of random antibodies created at the end of each iteration cycle. The rest of the parameter set contains one or more from the above presented stopping criterion.

4 Modeling Results

Our fuzzy modeling project is aimed at the creation of a rule-based system that can predict the number of students who will fail the Network Administration (NA) exam and therefore will enroll in the course again in the subsequent semester.

We chose a two-level solution. At the first level we determined the probability (PS) of passing the Network Administration exam for each student using a fuzzy rule based system and at the second level we calculated an average probability for the whole group of the students who enrolled for the NA course by

$$P_{av} = \frac{\sum_{i=1}^{N_{st}} P_i}{N_{st}}, \tag{9}$$

where N_{st} is the number of students enrolled in the current semester. Next, we calculated the estimated number of students who will pass the exam by

$$N_p = P_{av} \cdot N_{st} = \frac{\sum_{i=1}^{N_{st}} P_i}{N_{st}}. \tag{10}$$

In the course of the generation of the fuzzy model we took into consideration historical data of four past semesters with a total of 122 data records. Thirty records were taken out randomly for test purposes and the remaining 92 data rows were used for the creation and tuning of the rule base.

Based on the available data and previous experiences a rule base containing 21 rules was constructed by human experts. In case of each student we used four input values and one output which are presented below.

Input

- What was the result of the student's last (successful) Computer Networks I (CONI) exam?
  Variable: $N1R$, possible values: 2..5.
- Is this the 1st, 2nd, or 3rd enrollment of the student into the Network Administration course?
  Variable: NEN, possible values: 1..3.
- How many credits has the student earned up to now?
  Variable: NCr, possible values: SM, AV, BG

Output

- Probability of passing the exam
  Variable: PS, possible values: SM, AL, AH, BG

In the first three input dimensions we had crisp values which could be represented by singleton membership functions (see Fig. 3). In case of the fourth input dimension and the output dimension we created partitions with three or four linguistic values (see Figs. 3 and 4).

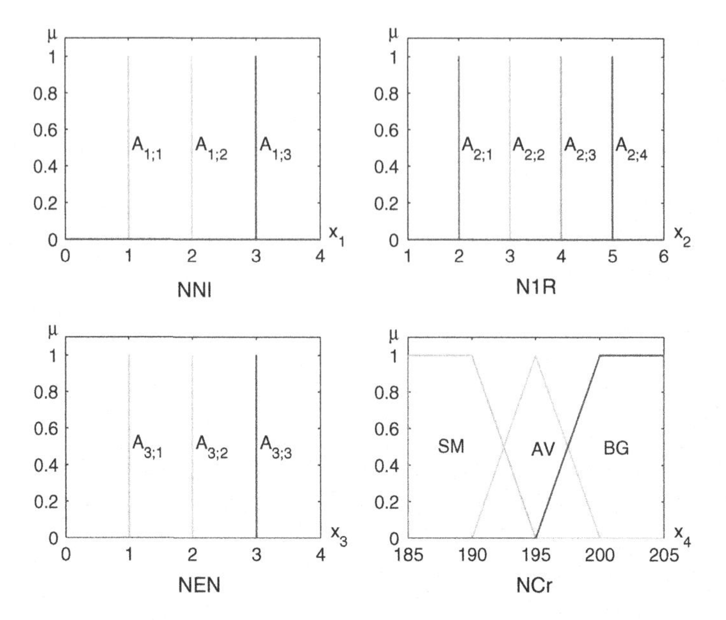

Fig. 3 Input partitions of the fuzzy system

Fig. 4 Output partition of the fuzzy system

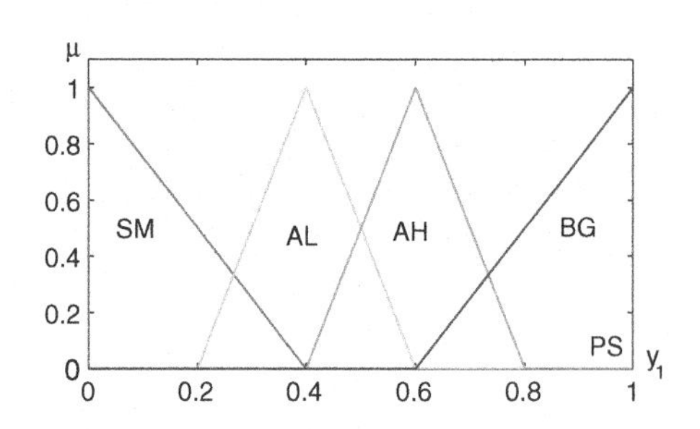

We applied the clonal selection based optimization algorithm presented in the previous section for the tuning of the rule base's parameters. The ordinal number of the consequent fuzzy sets in their partition was used as a parameter (values between 1..4) in the case of each rule (SM-0, AL-1, AH-2, BG-3).

We chose the following values for the parameters of the algorithm: $N = 30$, $n_R = 15, p = 0.5, n = 10, \beta = 0.4, \rho = 0.8, d = 10$. We used $n_g = 30$ as stopping criteria. The performance of the fuzzy system was measured by the formula

$$PI = 1 - \sqrt{\frac{\sum_{i=1}^{n_s} (y_i - \hat{y}_i)^2}{n_s}} \cdot \frac{1}{y_{max} - y_{min}}, \tag{11}$$

where n_s is the number of data points involved in the evaluation, y_{min} and y_{max} are the lower and upper bounds for the output values, y_i is the ith output value, and $\hat{y}_i$ is the ith estimated output value (calculated by the fuzzy system).

The initial fuzzy system with the rules created based on out previous experiences provided a performance of $PI = 0.73$ in the case of the training data and $PI = 0.74$ in the case of the test data set. The optimization slightly improved the performance of the system increasing it to $PI = 0.81$ in the case of the training data set and 0.76 in the case of the test data set.

5 Conclusions

In this chapter we reported the results of our efforts regarding the creation of a fuzzy system that can predict the number of students who will fail the Network Administration exam. The main purpose of this work was to support our department's resource allocation activity. Although an exact prediction cannot be made the results can be useful and our further research will concentrate on the improvements by taking into consideration a larger amount of historical data, and increasing the number of parameters included into the tuning process. An examination of other input values is also under consideration.

Acknowledgments This research was supported by the National Scientific Research Fund Grant OKTA K77809. The described work was carried out as part of the TÁMOP-4.2.2/B-10/1-2010-0008 project in the framework of the New Hungarian Development Plan. The realization of this project is supported by the European Union, co-financed by the European Social Fund.

References

1. Baranyi, P., Kóczy, L.T., Gedeon, T.D.: A generalized concept for fuzzy rule interpolation. IEEE Trans. Fuzzy Syst. **12**(6), 820–837 (2004)
2. Blažič, S., Škrjanc, I., Matko, D.: Globally stable direct fuzzy model reference adaptive control. Fuzzy Sets Syst. **139**(1), 3–33 (2003)
3. Botzheim, J., Hámori, B., Kóczy, L.T.: Extracting trapezoidal membership functions of a fuzzy rule system by bacterial algorithm. In: 7th Fuzzy Days, pp. 218–227. Springer, Dortmund (2001)
4. Brownlee, J.: Clonal selection algorithms. CIS Technical Report 070209A. Swinburne University of Technology, Melbourne, Australia (2007)
5. de Castro, L.N., von Zuben, F.J.: Artificial Immune Systems—Part I: Basic Theory and Applications. Department of Computer Engineering and Industrial Automation, School of Electrical and Computer Engineering, State University of Campinas, Brazil, TR DCA 01/99 (1999)
6. Chen, S.M., Ko, Y.K.: Fuzzy interpolative reasoning for sparse fuzzy rule-based systems based on α -cuts and transformations techniques. IEEE Trans. Fuzzy Syst. **16**(6), 1626–1648 (2008)
7. Detyniecki, M., Marsala, C., Rifqi, M.: Double-linear fuzzy interpolation method. In: IEEE International Conference on Fuzzy Systems (FUZZ: 2011), Taipei, pp. 455–462, 27–30 June 2011
8. Devasenapati, S.B., Ramachandran, K.I.: Hybrid fuzzy model based expert system for misfire detection in automobile engines. Int. J. Artif. Intell. **7**(A11), 47–62 (2011)
9. FRI Matlab ToolBox. http://fri.gamf.hu

10. Gál, L., Botzheim, J., Kóczy, L.T.: Advanced bacterial memetic algorithms. In: 1st Győr Symposium Computational Intelligence, pp. 57–60, (2008)
11. Huang, Z.H., Shen, Q.: Fuzzy interpolation with generalized representative values. In. Proceedings of the UK Workshop on Computational Intelligence, pp. 161–171 (2004)
12. Johanyák, Z.C.: Fuzzy rule interpolation based on subsethood values. In: Proceedings of 2010 IEEE International Conference on Systems Man, and Cybernetics (SMC 2010), pp. 2387–2393, 10–13 Oct 2010
13. Johanyák, Z.C.: Clonal selection based parameter optimization for sparse fuzzysys-tems. In. Proceedings of IEEE 16th International Conference on Intelligent Engineering Systems (INES: 2012), pp. 369–373, Lisbon, 13–15 June 2012
14. Johanyák, Z.C., Kovács, S.: Fuzzy rule interpolation by the least squares method. In: 7th International Symposium of Hungarian Researchers on Computational Intelligence (HUCI 2006), pp. 495–506, Budapest, ISBN 963 7154 54 X (2006)
15. Kovács, L.: Rule approximation in metric spaces. In: Proceedings of 8th IEEE International Symposium on Applied Machine Intelligence and Informatics SAMI 2010, pp. 49–52, Herl'any, Slovakia (2010)
16. Kovács, S.: Extending the fuzzy rule interpolation "FIVE" by fuzzy observation. advances in soft computing. In: Reusch, B. (ed.) Computational Intelligence, Theory and Applications, pp. 485–497. Springer, Germany (2006)
17. Perfilieva, I., Wrublova, M., Hodakova, P.: Fuzzy interpolation according to fuzzy and classical conditions. Acta Polytech. Hung. **7**(4), 39–55 (2010)
18. Precup, R.-E., Preitl, S., Faur, G.: PI predictive fuzzy controllers for electrical drive speed control: methods and software for stable development. Comput. Ind. **52**(3), 253–270 (2003)
19. Precup, R.-E., Preitl, S.E., Petriu, M., Tar, J.K., Tomescu, M.L., Pozna, C.: Generic two-degree-of-freedom linear and fuzzy controllers for integral processes. J. Franklin Inst. **346**(10), 980–1003 (2009)
20. Portik, T., Pokorádi, L.: Possibility of use of fuzzy logic in management. In: 16th Building Services. Mechanical and Building Industry days International Conference, pp. 353–360, Debrecen, Hungary, 14–15 Oct 2010
21. Shepard, D.: A two dimensional interpolation function for irregularly spaced data. In: Proceedings of the 23rd Annual International ACM SIGIR Conference, pp. 517–524 (1968)
22. Sinčák, P., Hric, M., Vaščák, J.: Neural networks classifiers based on membership function ARTMAP. In: Systematic Organisation of Information in Fuzzy System, Series: NATO Science Series. Subseries III: Computer and Systems Sciences, vol. 184, pp.321–333. IOS Press, Amsterdam (2003)
23. Tikk, D., Kóczy, L.T., Gedeon, T.D.: A survey on the universal approximation and its limits in soft computing techniques. Int. J. Approximate Reasoning **33**, 185–202 (2003)
24. Vass, Gy., Kalmár, L. Kóczy, L.T.: Extension of the fuzzy rule interpolation method. In: International Conference on Fuzzy Sets Theory and Applications, Liptovsky M. Czechoslovakia, pp. 1–6 (1992)
25. Vincze, D., Kovács, S.: Incremental rule base creation with fuzzy rule interpolation-based Q-learning. Stud Comput Intell Comput Intell Eng **313**, 191–203 (2010)

Optimization Tasks in the Conversion of Natural Language Texts into Function Calls

Péter Barabás and László Kovács

Abstract Natural language processing (NLP) is a well-known and increasingly more important area in human-computer interaction. Talking with a computer in human language is not very far away; for instance, there are automotive systems nowadays where people can control their car by voice, with some limitations. The goal of our research is to develop a natural language framework which will be used to extend existing systems with a natural language controlling capability. Our research focuses on the textual form of input; a proper speech recognizer or speech-to-text converter can produce textual commands and queries anytime. This chapter mainly deals with the optimization of algorithms in the most relevant modules of the framework: POS tagging and function mapping.

1 Introduction

Communication with most computer systems happens via user-friendly but not natural ways. People need to fill in forms, click buttons, or type instructions through a keyboard instead of giving commands in a natural language to the system. Natural language processing is an increasingly popular area of artificial intelligence for applications such as controlling robots, automotive systems, navigation systems, etc.

The main interaction form between human and computer is speech in spoken dialog systems (SDS). Three types of such systems can be distinguished: state-based [1–3], frame-based [4, 5] and agent-based [6].

P. Barabás(✉) · L. Kovács
Department of Information Technology, University of Miskolc-Egyetemváros, Miskolc 3515, Hungary
e-mail: barabas@iit.uni-miskolc.hu

L. Kovács
e-mail: kovacs@iit.uni-miskolc.hu

G. Bognár and T. Tóth (eds.), *Applied Information Science, Engineering and Technology*, Topics in Intelligent Engineering and Informatics 7, DOI: 10.1007/978-3-319-01919-2_3,

The state-based systems are the simplest and most commonly used. These kinds of dialog systems represent series of states. In each state the system asks the user for specific information. After every state is 'filled in', the system can generate an answer by several techniques: e.g., calling functions or running external applications. The algorithm for processing of input is based on recognition of well-defined words. State-based approaches are used for simple natural language processing (NLP) tasks.

In more complex cases, frame-based techniques are used instead of states, as in [4, 5]. A frame represents a task that has slots. A slot contains a piece of information which the system needs in order to complete the task. Several slots can be filled in at one time and the system can construct questions for empty slots. A slot can be marked as required or optional. If all required slots are filled, the system can complete the task and can generate the answer.

McTier [6] defines the agent-based system as a system where a frame is filled in cooperation with the user and a solution to the problem is sought together with the user. The system and user exchange knowledge and reasons about their own actions and beliefs to complete previously tasks.

Our goal is to define and implement a natural language framework using a frame-based dialog system that can be applied to robot control. Our framework processes text input that can be generated by a speech-to-text converter. The primary language supported in our system is Hungarian, which is very difficult to process because of its agglutinate property. The optimization of algorithms is also essential to produce an accurate and quick response system.

In Sect. 2 the text conversion process is introduced, describing the difficulties and special cases in each phase. A discussion of existing and proposed algorithms of POS tagging and its optimizations is included in Sect. 3. The other relevant process, function mapping, is covered in Sect. 4 together with its optimization solutions. The results and future work are summarized in the last section.

2 Text Conversion Process

The goal of text conversion is to produce a structured output from a text which can be interpreted by applications and controllable systems. Natural language text processing is a complex task which is divided into the following phases:

- Text cleaning, tokenization,
- Morphology analysis,
- Sentence analysis,
- Function mapping.

2.1 Text Cleaning, Tokenization

Input texts can be produced from several sources such as keyboards, speech-to-text converters, OCR systems, handwriting recognizers, etc. Each source type and converter can have specific faults and error rates. The input text should be normalized first so that as accurate as possible function mapping can be determined. It is important to note that the mistakes of the input text spin off the subsequent modules with decreasing output accuracy.

The input text should be parsed into sentences and each sentence should be split into words. There are several existing other tokenizers, for instance Stanford [7] or NLTK [8], which support mostly English language and several other widely spoken languages. There is a Hungarian toolkit for linguistic processing called "magyarlanc" [9] developed by the University of Szeged, which contains a sentence splitter, a tokenizer, a POS tagger and lemmatizer, stop word filtering and a dependency parser.

The result of tokenization is a set of words which needs to be analyzed further. The analysis consists of two phases: spell checking and named-entity recognition. There are two situations which the spell checker should be able to handle:

1. There is a mistyped word in the input which is in a dictionary of the language.
2. There is a mistyped word in the input which is not part of the dictionary.

In the first case only a context-sensitive spell checker can recognize if there is a wrong word in the sequence, in the second case a simple spell checker should already have detected the misspelled words. It may happen in both cases that several possible corrections are available but only one can fit the sentence. The use of an accurate context-sensitive spell checker like that in [10] is proposed, since in natural text processing human decisions should be minimized as much as possible.

The second phase should recognize the named entities such as numbers, places, companies, dates, points of interest, proper nouns, addresses, etc. in the text. The words belonging to named entities should be labelled, denoting that in later processing modules the word is to be handled in a special way, e.g. the morphology analyzer will not analyze these kinds of words.

The result of text normalization will contain the list of sentences consisting of spell-checked words labelled by the named-entity recognizer.

2.2 Morphology Analysis

In order to capture the meaning of the source text, the grammar structure of the text should be determined first. The grammar of the free text input can be modelled with a multi-layer architecture. Usually four levels are distinguished: the lowest level is the letter level, the next is the word level and the third level relates to the sentences. The highest level corresponds to the dialog or document level. The morpheme analyzer engines determine the grammatical structure of the words. At this level, the list of related suffixes, prefixes and the stem component will be discovered. These components can be assigned to the words as morpheme labels. In the case of languages like Hungarian, morphology analysis is quite a complex task because of their agglutinative property and the huge number of word forms. Hunmorph [11] is an open-source word analysis tool that contains a morphological analyzer and synthesizer.

The result of morphology analysis is the list of morpheme tags and the chain of suffixes with suffix codes, like in the case of the next word:

ablakokban = ablak [NOUN] + ok [PLUR] + ban [CAS(INE)]

In many cases a word can have several potential analyses. Similar to spell checking the proper analysis can be chosen only if the context of the current word is known. This task will be fulfilled in the sentence analysis module.

2.3 Sentence Analysis

Sentence analysis is the kernel module of text processing. It receives the morphologically labelled word sequence and has to produce an analysis tree. It can be stated that without prior knowledge about concepts belonging to the domain of the sentences, the sentence analysis cannot be accomplished accurately. For instance, if the predicate concept of the sentence is unknown, the other parts of speech cannot be bound and determined. Morpheme analysis helps a great deal in guessing the parts of speech for a sentence, but there is no definite mapping between most of the morpheme labels and parts of speech. Taking the sentence

A történelem könyv nagyon érdekes. (S1)
(The history book is very interesting.),

the morpheme analysis of sentence is the following:

A [ART] történelem [NOUN] könyv [NOUN]
nagyon[ADV,nagy[ADJ[CAS(SUE)]] érdekes[ADJ]

The parts of speech of sentence (1) should be:

- predicate: érdekes [ADJ]
  - predicate-adverbs of measure: nagyon [ADV]
- subject: könyv[NOUN]
  - subject-adjective: történelem [NOUN]

The problem is that it cannot be declared that the *noun* without suffix is always the subject or the predicate is always a *verb*. Certainly when the predicate is *verbal* instead *nominal*, the analysis is much easier. In the above case the following concepts of the domain are known:

1. There are objects in the school domain like books, pens, bags, cell phones, etc.
2. Categories of books can be history, grammar, computer science, etc.
3. Subjects like books can be interesting, boring, amusing, etc.
4. The properties can have absolute or relative measurements: much, more, less, best, least, etc.

Concept trees can be built from these statements which can be used in sentence analysis. The sentence analysis is a tree whose root node is the predicate element of the sentence. The hierarchy property of the tree allows the representation of compound structures in a sentence. In sentence analysis the order between nodes which are on the same level is arbitrary, but the order between levels is well-defined: the word belonging to a higher level node stands after the word of a lower level node.

The high time cost of algorithms and their low accuracy always cause optimization problems, therefore many existing POS tagger methods will be analyzed in Sect. 3.

2.4 Function Mapping

The goal of the last module in natural language text processing is to provide a format for sentence analysis which can easily be interpreted by the target applications or systems. For example, in order to call web services at the end of the process, we have to map the sentence analysis into a service URL and name-value pairs as parameters. This is called function mapping, where predefined functions are given and the concepts in sentence analysis have to be paired with a target function and its parameters.

First of all a function description model has to be built, which can also be a tree similar to sentence analysis. Figure 1 shows the nodes of function description.

The predicate of the sentence has a central role in function mapping also: it determines the target function to be called. Several predicates can be bound to a function if their meanings are similar. A function can usually have zero or more parameters. A parameter can be bound to a sentence analysis concept that has the defined part of a speech tag (constituent in Fig. 1). There can be such situations when several concepts exist in a sentence with the same part of speech tags. In this case a dependency can be defined in the parameter description between the concept and its parent concept in sentence analysis. For instance, there can be several adjectives in the sentence in different positions and the attached concept decides to which parameter they should be mapped. Let us take the following sentences:

Hány forintba kerül 10 euró?(How much Forint does 10 Euro cost?) (S2)

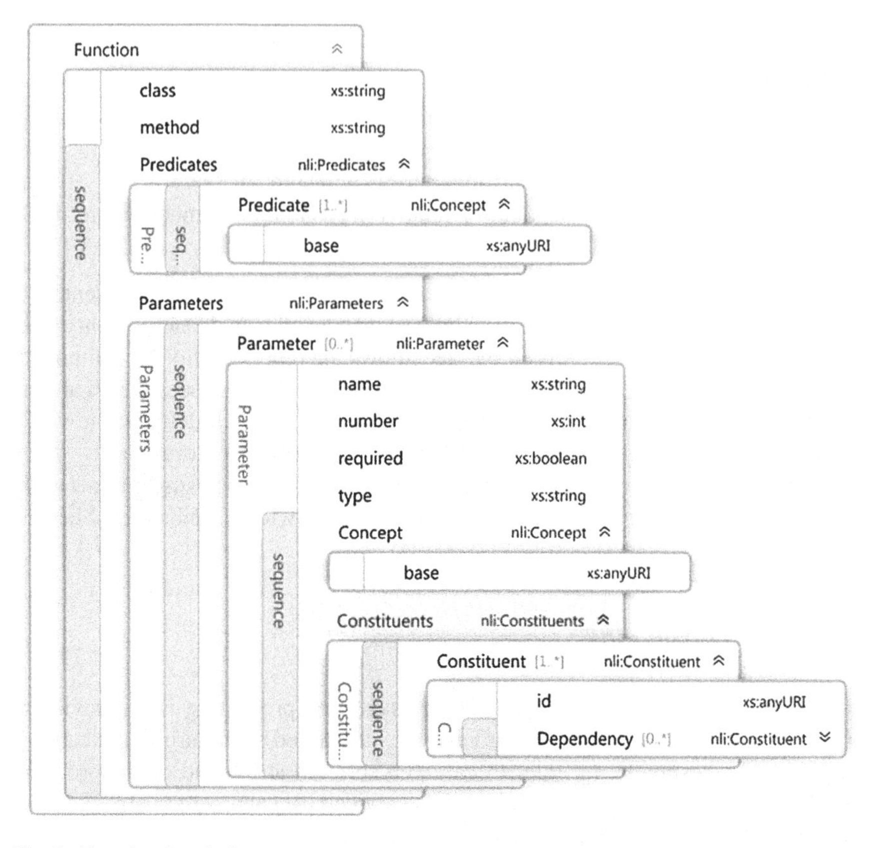

Fig. 1 Function description tree

Hány forint kerül 10 euróba?(How much Forint does 10 Euro cost?) (S3)

And we have a function:

cost(source)_currency, target_currency, source_count, target_count)

In case of S1, the function call is:

cost(EUR,HUF,10,?)

In case of S2, the function call is:

cost(HUF,EUR,?,10)

The source currency is the subject in both cases, while the target currency is the adverb. There is a problem here with the adjectives of quantities since they belong to the third and fourth parameters also. The parent concept designates their proper positions.

The mapping between the sentence analysis tree and the function description tree should be as accurate and fast as possible to be applied in real-time natural language controlling systems. For the sake of performance the previous properties function mapping algorithm should also be optimized, which is discussed in Sect. 4 in detail.

3 Optimization in POS Tagging

The morphological information alone is usually not enough to infer the semantic role of the word within the text. The sentence level structure encodes the required additional information to determine the semantic role. At the sentence level, the grammatical role determines the functional role of the word together with the relationships to the other words. A human sentence can be interpreted as a complex structure containing parts with different semantic grammatical roles. A typical sentence contains e.g. a predicate part and a subject component. These roles are the part of speech (POS) units of the sentence. Simple parsers usually use about 20 different roles while the more sophisticated models distinguish among more than 100 POS variants. The set of common POS roles includes among others the following items: NN (singular noun), NNS (plural noun), VB (verb base form), VBD (verb past tense), PN (personal pronoun), AT (article), IN (preposition), RB (adverb) or JJ (adjective) [12]. In the tagging process, the words of an input sentence will be assigned to a corresponding POS unit.

The main difficulty in the tagging process arises from the ambiguity of the words: a word may have different meanings and a word sequence may have different POS taggings. The following sentences give examples of these difficulties:

Boys [NN] move [VB] to [IN] a [AT] new [JJ] city [NN]

(There is) [VB] a [AT] move [NN] in [IN] (New York) [NN]

3.1 Markov POS Tagger

The usual approach in tagging is to consider the word sequence as a random Markov process. The Markov chain model [12] assumes that the state value at time point t depends only on the state value of the previous time point t-1. The second key assumption is that the probability of transition between any states s_1 and s_2 in time points t-1 and t is stationary, i.e. this probability is independent from time point t.,. This means that the probability of a tag sequence (tag denotes the POS attribute, the role of the word) can be given with the neighborhood probabilities as

$$p(t_{1..n}|w_{1..n}) = \frac{p(w_{1..n}|t_{1..n})p(t_{1..n})}{p(w_{1..n})} = \frac{\prod_i p(w_i|t_i)p(t_i|t_{i-1})}{p(w_{1..n})} \quad (1)$$

where the symbols are as follows:

$w_{1..n}$ word sequence between positions 1 and n
$t_{1..n}$ tag sequence between positions 1 and n
w_i word at positions i
t_i tag at positions i

Thus the most probable tag sequence belonging to a given input word sequence is calculated as

$$arg \max_t \{ \prod_i p(w_i|t_i)p(t_i|t_{i-1}) \} \quad (2)$$

The probability values between the neighboring states are estimated with the relative frequency values. An efficient implementation of tagging optimization is the Viterbi algorithm [13]. The Viterbi algorithm uses a dynamic programming approach to find the optimal tag sequence. It stores the cost value for every possible intermediate state in a matrix. An intermediate state is characterized with a time index and with a tag index. Based on the stationary assumption, the $p(t_i|t_{i-1})$ value is independent of the position, depending only on the tag values, i.e.

$$p(t_i|t_{i-1}) = p(t^k|t^l), \quad (3)$$

where t^k and t^l denote the corresponding tag values in the tag dictionary. The cost of the intermediate state with index *(i+1, j)* is calculated with

$$c(i+1, j) = \max_l \{ c(i, l)p(t^j|t^l)p(w_{w+1}|t^j) \} \quad (4)$$

where l denotes a tag index at time point i.

The HMM method belongs to the family of generative models where the joint distribution is the base formula to calculate the conditional probabilities. Another approach is represented by the discriminative model, which focuses directly on conditional probability. The Linear-chain Conditional Random Field (LCRF) [14] is an efficient alternative to the HMM method.

3.2 Linear-Chain Conditional Random Field

The main idea in LCRF is to segment the variables into smaller disjoint groups where each group is independent of the other groups. This process is called factorization. In LCRF, the conditional probability is calculated as [15]:

$$p(t_{1..n}|w_{1..n}) = \frac{1}{Z(w_{1..n})} \prod_i e^{\sum_k \Theta_k f_k(t_i, t_{i-1}, w_i)} \quad (5)$$

where *Z()* is normalization function, *f()* is the feature function and θ is the parameter vector. The domain of feature engineering refers to the selection of appropriate feature functions. The number of feature vectors may be very large in practical applications; for example in [16] about 3.5 million features were used. For a given feature and vector set, the Viterbi algorithm mentioned can be used to determine the most probable tag sequence.

In LCRF, a separate step is the selection of the appropriate parameter vector. For calculation of the maximum likelihood feature vector, numerical optimization is used. The proposed method usually optimizes the conditional log likelihood value. To avoid overfitting, the following regularized log likelihood is used for parameter optimization:

$$\epsilon(\Theta) = \sum_j \sum_i \sum_k \Theta_k f_k(t_i^j, t_{i-1}^j, w_i^j) - \sum_j \log Z(w_{1..n}^j) - \sum_k \frac{\Theta^2}{2\nu} \quad (6)$$

where the superscript *j* denotes the training data index and ν denotes a penalty factor. At the optimum parameter value, the derivate of the objective function is equal to zero. The solution of the system of equations

$$\sum_j \sum_i f_k(t_i^j, t_{i-1}^j, w_i^j) - \sum_j \sum_i \sum_{t,t'} f_k(t, t', w_i^j) p(t, t', w_i^j) - \frac{\Theta}{\nu} \quad (7)$$

yields the optimum parameter vector.

The main shortcoming of this model is that it considers a sentence as a linear chain. Classic grammar models like TAG (tree adjoining grammar) are based on this approach. The TAG formalism proposed by [17] defines initial and auxiliary trees. A tree encodes the set of possible sequences, where an element of the sequence can be replaced with other sub-sequences (adjunction operation). For languages with a strict word order, the sequence model is a good approach. However, the analysis of the sentence structures in the Hungarian language shows that sequence orientation is not the perfect model for languages with no dominant word order. In these languages, all permutations of the words may be valid. An explicit encoding of all possible permutations would result in intractable grammar trees. The grammar model in [21] unifies the features of contextual and tree-adjoining grammars using the standard structure based on tree components. To provide a more suitable formalism, a graph oriented model is proposed in our system.

Another key problem in tagging operation is to find the correct segment boundaries. Words belonging to different segments may have the same grammatical properties. In the sentence

Tegnap telefonált Peti Zoli barátjának. (S4)
(Peter phoned yesterday Zoli's friend)

After the morpheme analysis, both words Peti and Zoli will be labeled NN, thus both are candidate subjects of the sentence. The correct segmentation should find the following segments within the sentence:

- predicate : telefonált
- subject : Peti
- dat_object : Zoli barátjának
- adverb : tegnap

The segments may be of simplex structure or of complex structure: e.g., in a higher order statement a sub-statement can be used as a POS unit:

Mondtam neked, hogy gyere pontosan. (I told you to come on time.)

This example also shows the fact that sometimes some POS roles are implicitly given, and have no explicit word representative. The subject I is encoded in the inflection of the verb. Thus the POS segment structure of the sentence is:

- predicate : mondtam (past)
- subject : én (hidden , implicit)
- dat_object : neked
- object : hogy gyere pontosan
  - object-predicate : gyere
  - object subject : te (hidden, implicit)
  - object-adverb : gyorsan

The management of free word-order was also addressed in some previous grammar models. One important group of these kinds of models is the family of dependency grammars. Dependency grammar (DG) [18] is based on modeling the dependency between the different POS units: e.g. a predicate unit requires a subject and an object POS unit. The dependency is an asymmetric relationship, the head corresponding to the independent part of the relationship. A similar approach is presented in word grammar [19], which is based on dependency grammar. In this model, language can be represented by a network of propositions. This approach can be applied, in neurolinguistics domains, among others. There are some other approaches in the literature to deal with free word order. One of the proposals is the finite automata with translucent letters [22, 23].

3.3 Proposed Tagging Method

The semantic structure of sentences is given with a multi-level graph. In the graph model a graph is given as

$$G = \{l, K, G_N, G_D, G_P\},$$

where G_N denotes the set of nodes. A node may be either a simple node or a graph. There is a special node called kernel node K. The kernel node is always the source node of the dependency relationship. The symbol l is for the identifier label of the graph. Every graph corresponds to a segment of the sentence. Within the segment, there is a dependency relationship between the kernel node and the other nodes. The set of edges G_D corresponds to the dependency relationship between the items. The meaning of dependency is the same as it is defined in the dependency grammar [18]. The relationship $t_1 \longrightarrow t_2$ is met if POS t2 can occur in a sentence only if POS t1 occurs in the same sentence.

$$G_D = \{K \rightarrow L \,|L \in G_N\} \tag{8}$$

In the graph unit, every kernel node depends only on the kernel. Each node in the graph is given by a set of attribute tuples. An attribute tuple includes the following elements:

- a role name of the node,
- morpheme tags of the corresponding words (A_m),
- probability of a given morpheme tag set (A_p),
- flag to denote whether the node, role is optional or not (A_o).

The set G_P describes the precedence relationship between the elements of $\{G_N \cup K\}$. The set G_P may be empty, denoting that no word order rule can be discovered, and every word order is valid. A probability value is assigned to every element of G_P.

For the sample sentence (S4) the morpheme analyzer returns the following morpheme structure:

tegnap : [FN] + [NOM] | [HA]
telefonált : [IGE] + [MIB] + [e3]
Zoli : [FN] + [NOM]
Peti :[FN] + [NOM]
barátjának : [FN] + [PSe3] + [DAT]

The symbol [x] denotes here a morpheme tag of the word. In the example, the following symbols are used: FN(noun), HA(adverb), IGE(verb), DAT(dative). Based on this list, a graph model can be generated as it is shown in Fig. 2. The solid line corresponds to the dependency relationship, while the dashed line shows the precedence relationship among the graph nodes and the dotted line shows the precedence

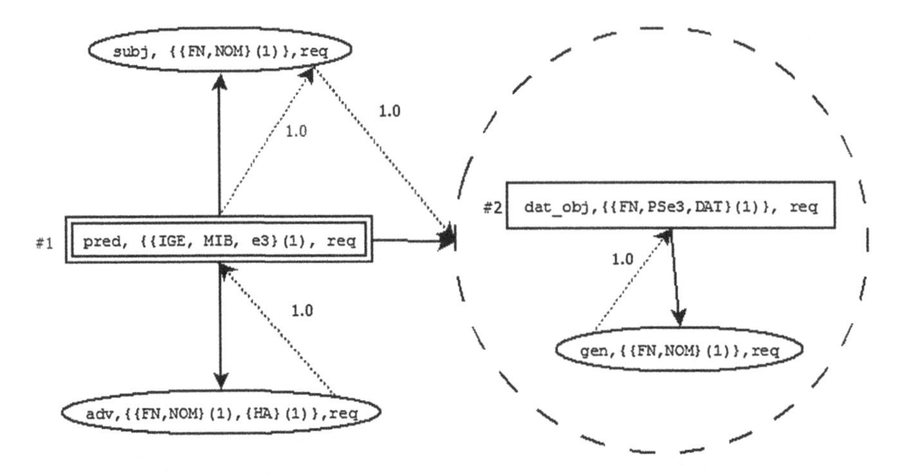

Fig. 2 Sample sentence graph for (S4)

relationship among the graph nodes. The positive numbers labels are the probability values of the relationship.

The graph structure contains the segmentation structure of the sentence. The kernel node of the main segment is a predicate node (double-line border) and the kernel of the sub-segment is a non-predicate word (single-line border).

In the graph model, the words assigned to the graph nodes correspond to words w_i of the sentence. The role name of the edges pointing to the node denotes the tag name t_i. The morpheme structure is a set of atomic morpheme units. The conditional probability $p(a|r)$ for each morpheme unit can be calculated with the sum of the corresponding tuple probabilities. Based on the graph, the tag role value can be estimated both from the morpheme structure and the precedence relationship.

For the morpheme-based probability p_m, we use the Bayes formula:

$$p_m(r|w) = \left\{ \frac{p(r) \prod_{a \in w} p(a|r)}{p(w)} \right\}. \tag{9}$$

When the independency assumption is known, the morpheme probability for a tag sequence is equal to the product of the probabilities of the components:

$$p_m(r_{1..n}|w_{1..n}) = \prod_i p_m(r_i|w_i) \tag{10}$$

The precedence probability for the tag sequence is given as

$$p_p(r_{1..n}|w_{1..n}) = \prod_i p_p(r_i r_j) \tag{11}$$

where p_p comes from the graph model directly. The sum of the morpheme and of the precedence probabilities is defined as the probability of a tag sequence:

$$p(r_{1..n}|w_{1..n}) = p_m(r_{1..n}|w_{1..n}) + p_p(r_{1..n}|w_{1..n}). \tag{12}$$

For a given word sequence, the algorithm selects the tag sequence with the maximum weight:

$$p(r_{1..n}|w_{1..n}) = argmax_r\{p(r_{1..n}|w_{1..n})\}. \tag{13}$$

In order to reduce the cost of a brute force search operation, a dynamic optimization method can be used similar to the Viterbi algorithm.

4 Optimization in Function Mapping

The difficulty of function mapping is caused by ambiguous mapping between sentence analysis and function signatures (*FS*). This can occur when there are concepts in the domain that can fit into many *FS*, predicates are not assigned to an individual *FS*, there are missing parameters in the sentence, and similar problems arise that make mapping much harder. A mapping between a sentence analysis tree and *FS* can be defined as:

$$S \rightarrow f_i | F = \{f_i\}, \tag{14}$$

where S is the sentence analysis tree and f_i is the ith function signature from the set of function signatures F. Let us define a fitness function Φ which calculates the correspondence value between sentence S and a given function signature f_i:

$$\phi_i : (S, f_i) \rightarrow R \tag{15}$$

The goal of function mapping is to find the winner function Φ_w that the has highest fitness value:

$$w = \arg\max_i\{\phi_i\} \tag{16}$$

The proper fitness function is the key component of this module, where the following rules have to be taken into consideration:

1. A sentence analysis tree has only one predicate and can have several other concepts. (We work only with simple, extended sentences.)
2. A predicate can be assigned to one or more *FS*.
3. A concept of the analysis tree should be mapped to one of the parameters of *FS*.
4. A concept can be mapped only to one parameter of *FS* at one time.
5. A concept may not belong to any parameter.

6. An *FS* can have required and optional parameters, where finding the required parameters is more important. If a required parameter is missing, the method belonging to *FS* cannot be executed.
7. If a parameter has a dependency node, only the sentence analysis node with the proper parent can be mapped.

Let us define a sentence analysis tree as follows:

$$S = (C, \rightarrow), C = \{P\} \cup \{c_i\}, \rightarrow \subseteq C \times C \tag{17}$$

where P is the predicate concept and c_i is a non-predicate concept, and $\longrightarrow$ is the link between concepts in the tree. The following statements are valid for analysis tree:

$$\forall c_i \in C \Rightarrow P \rightarrow c_i \; and \; \forall c_i, c_j \in C : c_i \rightarrow c_j \Rightarrow c_j \nrightarrow c_i. \tag{18}$$

The set of *FS* are defined in (19):

$$F = \{f_i\}, \;\; f_i = (\Gamma, \Pi), \;\; \Gamma = \{\gamma_i\}(i = 1..n)\Pi = \{\pi_j\}(j = 1..m) \tag{19}$$

where Γ is the set of predicates and Π is the set of parameters.

A parameter defined in (24) has some attributes like name, position, type and mandatory. In constructing the fitness function only the mandatory property can have a significant role; the others are used in preparing the execution of the method belonging to the given *FS*.

$$\pi_i = (m, \Theta), (m \in \{0, 1\}) \; \Theta = \{\theta_i\}(i = 1..n) \tag{20}$$

Θ means the constituents of the parameter. A constituent can have zero or more dependencies:

$$\theta = \{\delta_i\}(i = 1..n). \tag{21}$$

Fitness value points can be given for an *FS* in the following scenarios:

1. The predicate of a sentence is found among predicates of the given *FS*.
2. A non-predicate concept can be mapped to any of parameters of the given *FS*.
   (a) If a parameter has dependency, the fitness score can be applied only if the concept and its parent are a fit.
3. A higher score will be applied if the actual parameter is required.

Since a predicate can be matched for many functions, relevance is an important factor in choosing the proper function. The fewer number of functions match a given predicate, the higher the predicate relevance is. A tf-idf measure [12] can be used to determine the relevance of a predicate:

$$tf(P, f_i) = freq(P, f_i), \tag{22}$$

$$idf(P, F) = \frac{\log([f_i])}{|\{f_i : P(f_i)\}|}, \tag{23}$$

$$tfidf(P, f_i, F) = tf(P, f_i) \times idf(P, F). \tag{24}$$

In the mapping score system we have three different scores: predicate relevance (x), the number of required parameters found (y), and the number of parameters found (z).

The resultant value can be computed with weighted sum of the above scores:

$$\Phi_i = \alpha x_i + \beta y_i + \delta z_i \tag{25}$$

where the weight factors are the input parameters of the algorithm.

5 Results and Future Works

In our research three prototype applications were developed for supporting the Hungarian language in natural language controlling: a navigation application, a currency query application and a robot controlling application. The navigation application (Fig. 3) can retrieve distances between cities or places, and uses Google Maps web services to find points of interests, etc. The currency query application is a simple question-answer application; we can query information about currency exchange rates today or in the past. It is also possible to query exchange rates in a period of

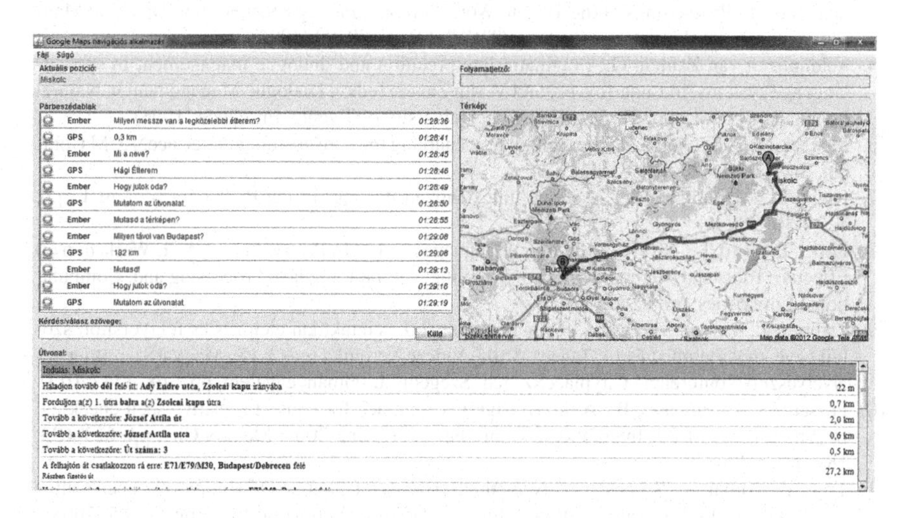

Fig. 3 Prototype applications

time which is displayed in a chart. The third application controls a humanoid Nao robot with simple commands.

Concepts of three different domains had to be built and the proper function declarations also had to be written. The operability of applications proves the applicability of the NLP framework.

In the future we want to examine the language-dependency of the framework and to make proper modifications in the engine to support other languages besides Hungarian with ease.

Another future plan is to re-factor and organize the source codes of the NLP framework for use in research projects by other NLP researchers.

Acknowledgments The described work was carried out as part of the TÁMOP-4.2.2/B-10/1-2010-0008 project in the framework of the New Hungarian Development Plan. The realization of this project is supported by the European Union, co-financed by the European Social Fund.

References

1. Aust, H., Oerder, M.: Dialogue control in automatic inquiry systems. In: Dalsgaard, P., Larsen, L., Boves, L., Thomsen, I. (eds.) Proceedings of the ESCA Workshop on Spoken Dialogue Systems, pp. 121–124. Vigso, Denmark (1995)
2. McTear, M.: Spoken Dialogue Technology: Enabling the Conversational User Interface, Distributed at the DUS/ELSNET Bullet Course on Designing and Testing Spoken Dialogue Systems, Springer, New York (1997)
3. McTear, M.: Modelling spoken dialogues with state transition diagrams: experiences of the CSLU toolkit. In Proceedings of the International Conference on Spoken Language Processing, Australian Speech Science and Technology Association, Incorporated, Sydney, Australia **4**, pp. 1223–1226 (1998)
4. Hulstijn, J., Steetskamp, R., ter Doest, H., van de Burgt, S., Nijholt, A.: Topics in SCHISMA dialogues, In Proceedings of the Twente Workshop on Language Technology: Dialogue Management in Natural Language Systems (TWLT 11) pp. 89–99 (1996)
5. Veldhuijzen van Zanten, G.: Pragmatic interpretation and dialogue management in spoken-language systems, In Luperfoy, N., and V. van Zanten (eds.), Dialogue Management in Natural Language Systems, pp. 81–88, (TWLT11) University of Twente, Enschede (1996)
6. McTear, M.F.: Spoken dialogue technology: enabling the conversational interface. ACM Comput. Surv. **34**(1), 90–169 (2002)
7. Klein, D., Manning, C.D.: Accurate unlexicalized parsing. In the Proceedings of ACL, pp. 423–430 (2003)
8. Loper, E., Bird, S.: NLTK: The Natural language toolkit. In Proceeding of ACL Workshop on Effective Tools and Methodologies for Teaching Natural Language Processing and, Computational Linguistics, pp. 62–69 (2002)
9. Zsibrita, J., Vincze, V., Farkas, R.: magyarlanc 2.0: szintaktikai elemzés és felgyorsított szófaji egyé. egyértelműsítés.In: Tanács Attila, Vincze Veronika (eds.): IX. MagyarSzámítógépes Nyelvészeti Konferencia.CityplaceSzeged, Szegedi Tu-dományegyetem, pp. 368–374 (2013)
10. Fossati, D., Di Eugenio, B.: A mixed trigrams approach for context sensitive spell checking. In: Proceedings of 8th International Conference, CICLing 2007, Mexico City, pp. 623–633, 18–24 Feb 2007
11. Trón, V., Kornai, A., Szepesi, Gy., Németh, L., Halácsy, P., Varga, D.: Hunmorph: open source word analysis. In: Software '05 Proceedings of the Workshop on Software pp. 77–85 (2005)

12. Manning, C., Schütze, H.: Foundations of Statistical Natural Language Processing, MIT Publisher, Cambridge (1999)
13. Freitag, D., McCallum, A.: Information extraction with HMM structures learned by stochastic optimization. In Proceeding of AAAI (2000)
14. Lafferty, J.D., McCallum, A., Pereira, F.C.N.: Conditional random fields: probabilistic models for segmenting and labeling sequence data. In: Proceedings of the Eighteenth International Conference on Machine Learning (ICML 2001), pp. 282–289, Morgan Kaufmann Publishers, Burlington (2001)
15. Sutton, C., McCallum, A.: An Introduction to Conditional Random Fields. Now Publishers, Foundation and Trends in Machine Learning (2012)
16. Sha, F., Pereira, F.: Shallow parsing with conditional random fields. In: Proceeding of NAACL '03 Proceedings of the 2003 Conference of the North American Chapter of the Association for Computational Linguistics on Human Language Technology, pp. 134–141 (2003)
17. Joshi, A., Levy, L., Takahashi, M.: Tree adjunct grammars. J. Comput. Syst. Sci. **10**(1), 136–163 (1975)
18. Tesniere, L.: Elements de Syntaxe Structural, Publisher Kliencksieck, Paris (1959)
19. Hudson, R.: An Introduction to Word Grammar, Cambridge University Press, Cambridge (2010)
20. Paun, G.: Marcus Contextual Grammars, Kluwer Academic Publisher, Dordrecht (1997)
21. Dedin, A., Nagy, B.: Computing trees with contextual hypergraph grammars. In Proceeding of ForLing, pp. 39–53 (2007)
22. Nagy, B., Ottó, F.: Finite state acceptors with translucent letters. In: Proceeding of ICAART, pp. 3–31 (2011)
23. Nagy, B., Kovács, L.: Linguistic applications of finite automata with translucent letters. In: Proceeding of ICAART, pp. 461–469 (2013)

Pattern Distillation in Grammar Induction Methods

Zsolt Tóth and László Kovács

Abstract The rule extraction phase plays a very important role in Context-Free Grammar induction systems. The mining of frequent patterns and rules is a costly task. The preprocessing of the training set provides a way to make the related methods more efficient. Apriori and FP-Growth algorithms are the standard methods for determination of frequent itemsets. Two novel methods are presented for pattern mining in this chapter. The first one is based on extended regular expressions with multiplicity approach. The second method is based on the theory of concept lattices.

1 Introduction

Grammar Induction is an actively investigated area of knowledge engineering. It is widely used in many application areas such as understanding the structure of unknown languages and the discovery of the internal structures of unknown symbol sequences. There are many papers in the literature about its application fields as well as induction algorithms, a review can be found in [2]. The proposed methods are often tailored to a solution for a specific task. There are only a few works in the literature which aim at comparing the existing methods.

This chapter focuses on the pattern extraction phase of Context-Free Grammar Induction Systems. Natural languages can be modelled efficiently with CFG because CFG is able to describe some common behaviour of natural languages. But there are some structures of the natural languages which cannot be modelled by CFG. Pattern discovery is the first phase of CFG induction methods. Thus the efficiency and functionality of pattern recognition process depends significantly on the efficiency

Z. Tóth (✉) · L. Kovács
Department of Information Technology, University of Miskolc,
Miskolc Egyetemváros, 3515, Hungary
e-mail: tothzs@iit.uni-miksolc.hu

G. Bognár and T. Tóth (eds.), *Applied Information Science, Engineering and Technology*, Topics in Intelligent Engineering and Informatics 7, DOI: 10.1007/978-3-319-01919-2_4,

and functionality of the pattern distillation phase. Despite the great efforts made so far, no general solution to inducing CFG has been identified.

The chapter presents two novel approaches which are based on mining of frequent regular expressions and on the generalization mechanism of concept lattice. The reduction of the training set is an internal step which could make the induction methods more efficient. Pattern extraction methods remove or substitute the less frequent items; thus the relevant patterns can be determined at lower cost. The reduced training set contains only frequent patterns.These methods also support the implementation of an iterative algorithm which realizes incremental learning. The Apriori and FP-Growth algorithms were proposed for frequent itemset mining solutions in the reduction phase, because there are some common aspects in grammar inference and association rule extraction. Moreover a novel filtering method has been proposed which is based on the multiplicity operators of Regular Expressions. This method replaces the rare words of the corpus with well defined symbols, so that the filtered training set contains only frequent words and the novel special symbols. The proposed reduction method eliminates those sequences which have no significant role in the discovery of frequent symbol sequences.

The second approach is based on the theory of concept lattices. This algorithm takes a sequence as an object and the contained subsequences as attributes. This algorithm calculates the common parts and unique parts for each pair of samples. The common subsequence is the intersection of the two operands. The unique parts correspond to the sets of interchangeable symbols. In the chapter, a theoretically approximation is given for the cost function of the pattern distillation process. These proposed methods are implemented with a novel CFG induction algorithm.

2 Grammar Induction

Grammar Induction is a process, which generates a grammar description related to the sentences of the training set or corpus. An important foundation of Grammar Induction is the theory of formal grammars which is a mathematical formalism to process languages given with symbol sequences. A formal language is defined as a finite set of sentences over a finite set of symbols $\mathcal{L} \subseteq \Sigma^*$, called alphabet Σ. The language can be defined in several ways: by enumeration of the sentences or by a generative grammar. A grammar is defined as $\mathcal{G} = \langle \mathcal{T}, \mathcal{N}, \mathcal{P}, \mathcal{S} \rangle$. Where $\mathcal{T} = \{a, b, c, \dots\}$ denotes the finite set of terminal symbols, $\mathcal{N} = \{A, B, C, \dots\}$ denotes the finite set of non-terminal symbols and $\mathcal{T} \cap \mathcal{N} = \emptyset$. The difference between terminal and non-terminal symbols is that the non-terminal symbols can be replaced by a sequence of symbols. This replacement is defined by the set of production rules denoted by $\mathcal{P} = \{\alpha \rightarrow \beta\}$, where $\alpha \in \{\mathcal{T} \cup \mathcal{N}\}^* \{\mathcal{N}\} \{\mathcal{T} \cup \mathcal{N}\}^*$ i.e. α contains at least one non-terminal symbol and $\beta \in \{\mathcal{T} \cup \mathcal{N}\}^*$. The valid sentences of the $\mathcal{L}$ can be deduced in finite steps from a non-terminal symbol $S \in \mathcal{S}$ which is called sentence symbol where $\mathcal{S} \subseteq \mathcal{N}$.

Table 1 Chomskian hierarchy

No.	Grammar	Implementation	Rule type
3.	Recursive	Turing machine	$\alpha \rightarrow \beta$
2.	Context-sensitive	Linear bounded automaton	$\beta\alpha\omega \rightarrow \beta A\omega$
1.	Context-free	Push down automaton	$A \rightarrow \alpha$
0.	Regular	Finite automaton	$A \rightarrow a \mid aB \mid Ba$

Grammars can be categorised based on the complexity of their production rules. Chomskian hierarchy is the best known classification of grammars which distinguishes four categories (see Table 1).

The chapter focuses on the induction of a subset of formal grammars called Context-Free Grammar. CFG is a good approximation of both natural and artificial languages. There are many efficient algorithms to process CFG such as CYK, TBL [7, 8] or ADIOS [9] algorithms. The rules of CFG have the form $A \rightarrow \alpha$, i.e. a non-terminal symbol can be replaced with a sequence of terminal and non-terminal symbols. The sentences contain only words which are terminal symbols and cannot be replaced. This simple form of the production rule makes efficient algorithms available, for example the CYK parsing algorithm solves the membership problem in $O(n^3)$ time where n is the length of the sentence. CFG is powerful enough to model some common phenomena of natural languages, but not all of them. With a simple example it can be shown that natural languages are more complex than regular languages. The sample sentence has the following form: "A man who heard something (from a man who heard... and it was right,) and it was right". This kind of sentences are often represented formally as an abstract language $\mathcal{L} = \{0^n 1^n\}$. $\mathcal{L} = \{0^n 1^n\}$ can be described as $\mathcal{G}_{\mathcal{L}} = \langle\{0, 1\}, \{A\}, \{A \rightarrow 0A1\}, \{A\}\rangle$ in CFG. However in some natural languages such as Spanish and Russian the form of a word depends on its context. So CFG gives only a good approximation of natural languages, but it cannot cover each language properly.

A formal language can be given with an enumeration of the valid sentences or with a grammar description. The enumeration is extremely costly and it is often impossible. There are two ways to generate a grammar description for a language: a manual way and an automatic way. The manual generation requires an expert who has a profound knowledge of the language and is also costly, time-consuming task. In addition there are many languages which have no experts, e.g. the artificial languages of chemical or biological samples. Although, there are many proposals in the literature [2], the automatic induction of CFG is still an open question and it is an NP-hard problem.

The literature on CFG inference includes many methods based on different approaches and using various heuristics to reduce, compress or produce a grammar description. The efficiency of the induction method depends on both the internal representation of the samples, the data structures and the applied heuristics. Three phases can be distinguished in CFG induction methods: an initialization, a rule extraction and a generalization phase. In the first phase, the training samples are converted into an internal representation which could be a table-like data structure, a directed graph or a list of strings or etc. The representation affects the entire process, because the different data structures support various algorithms. Moreover the same representation form can be used with different inference methods. For example the TBL algorithm [7, 8] and the Inductive CYK [6] use the same table-like representation but, their applied heuristics are different. In this chapter the initial phase and the internal representation are emphasized and analysed based on some well-known algorithms.

2.1 Investigation of the TBL Algorithm and Inductive CYK Algorithm

The TBL [7, 8], Improved TBL [5, 12] and the Inductive CYK [6] algorithms use a table-like data structure similar to the parse table of the CYK parsing algorithm. The CYK algorithm uses a table-like data structure whose cells contain sets of non-terminal symbols. A n-length sentence ω is a sequence of words denoted by $\omega_1\omega_2\cdots\omega_n$. A non-terminal A symbol in the (i, j) cell of the matrix means that the $\omega_i\cdots\omega_{i+j}$ subsequence of the sentence can be deducted from non-terminal A. The output of the induction algorithm is a grammar in Chomskian Normal Form. The data structure both encodes the non-terminal symbols and the corresponding rules. The CYK is a bottom-up parsing algorithm and it fills the matrix iteratively. The construction of the table requires $O(n^3)$ operation steps where n is the length of the sentence, so it is a polynomial algorithm. The TBL and Inductive CYK algorithm are based on a similar table-like data representation.

The TBL algorithm uses a lower-triangle matrix whose cells contain arrays of non-terminal symbols. It is used to generate primitive grammars which contain each possible deduction tree with each sentence. The data structure of the CYK algorithm encodes that, a given part of the sentence which can be deducted from the non-terminal symbol. While the data structure of the TBL and the Improved TBL encodes the way of the deduction as well. The modified data structure contains arrays of non-terminal symbols instead of sets. The position in the array determines unambiguously the tail of the rule.

Table 2 shows a tabular representation of the sentence *Peter likes football too*. The complete primitive grammar can be determined from this table. Terminal symbols are in the bottom line of the table, non-terminal symbols are in the rest of the table. The rules are encoded by position e.g. $X_{100} \rightarrow X_{000}X_{010}$, $X_{200} \rightarrow X_{000}X_{110}$, $X_{201} \rightarrow X_{100}X_{010}$. From Table 2 the following deduction can be determined $X_{200} \rightarrow X_{000}X_{110} \rightarrow Peter X_{010}X_{020} \rightarrow$ "Peter likes football". The sentence symbols are in the top of the matrix $\{X_{300}, X_{301}, X_{302}\}$. Thus the whole sentence can be represented by one of the following non-terminal symbols: X_{300}, X_{301} and X_{302}.

The TBL and ITBL algorithms generate primitive grammar for each positive sample then merge them into an aggregated grammar in the distillation phase. It describes exactly each positive sentences and no more, so it does not accept any sentences

Table 2 T(Peter likes football too)

3	$[X_{300}X_{301}X_{302}]$			
2	$[X_{200}X_{201}]$	$[X_{210}X_{211}]$		
1	$[X_{100}]$	$[X_{110}]$	$[X_{120}]$	
0	$[X_{000}]$	$[X_{010}]$	$[X_{020}]$	$[X_{030}]$
	0	1	2	3
ω	Peter	likes	football	too

which were not in the training set. In the generalization phase the non-terminal symbols are clustered by a genetic algorithm which determines the equivalence classes of the symbols. This phase of the algorithm is the most costly one and the TBL and ITBL algorithms only differ in the genetic algorithm applied.

The Inductive CYK algorithm is based on the CYK parsing strategy and uses CYK parsing method to determine a subset of the grammar $\mathcal{G}_\omega$. It is the smallest subset of the terminal and non-terminal symbols, production rules and sentences symbols which are related to the training sample. The algorithm modifies only $\mathcal{G}_\omega$, so the modifications remain locally. It works iteratively and parses each positive sample with the grammar. If the sentence can be deducted from the grammar it takes the next sample. Otherwise it adds stochastically new non-terminal symbols to the parse table and extends the rule set of grammar $\mathcal{G}_\omega$, until $\mathcal{G}_\omega$ does not accept the sentence. The technique of backtracking is applied in order to avoid the learning of negative sentences. Thus the Inductive CYK algorithm applies a different approach and heuristic methods on the same representation as TBL and ITBL algorithms.

2.2 ADIOS Algorithm

The ADIOS algorithm [9] is an unsupervised Probabilistic CFG induction method which requires only positive samples. It builds a directed graph from the training set and extracts the patterns as frequent paths. Constant prefix (*begin*) and suffix (*end*) word are added to each sentences, and thus each sentence cam be considered as a *begin* to *end* path in the graph. The graph representation is very compact because there is one node for each word, thus the total number of the nodes is $\mid \mathcal{T} \mid +2$ due of the *begin*, *end* symbols. Multiple edges are used to distinguish each long path. Because the graph is directed each node has a fan-in and a fan-out value and it is the base of pattern extraction. A path in the graph is denoted by $e_i \rightarrow e_{i+1} \rightarrow \cdots \rightarrow e_j$ or $e_i \rightarrow e_j$.

The candidate patterns are the paths of the graph. Four values are calculated for each candidate path (e_i, e_j) which are the following:

Right moving ratio $P_R(e_1, e_n) = \frac{|e_1 \rightarrow e_2 \rightarrow \cdots \rightarrow e_{n-1} \rightarrow e_n|}{|e_1 \rightarrow e_2 \rightarrow \cdots \rightarrow e_{n-1}|}$

Left moving ratio $P_L(e_1, e_n) = \frac{|e_1 \rightarrow e_2 \rightarrow \cdots \rightarrow e_{n-1} \rightarrow e_n|}{|e_2 \rightarrow \cdots \rightarrow e_{n-1} \rightarrow e_n|}$

Right decrease ratio $D_R(e_1, e_n) = \frac{P_R(e_1, e_n)}{P_R(e_1, e_{n-1})}$

Left decrease ratio $D_L(e_1, e_n) = \frac{P_L(e_1, e_n)}{P_L(e_2, e_n)}$

These values describe the significance of the given path compared with others paths. The best candidate will be added to the graph as a new node which denotes the sequence of the nodes of the pattern. The algorithm works iteratively so the graph grows during the process with the new nodes. In each iteration the graph is refreshed thus the graph is always accurate.

3 Training Set Preprocessing

The grammar induction for complex languages is a very costly task due to the huge vocabulary and complex grammatical structures. A reduction of the training set is proposed, because the preprocessing step can reduce the execution cost of the induction process. An already filtered training set could contain only the relevant patterns. It also provides a way to develop an iterative algorithm because the rate of the reduction has an effect on the learning process. The increase of the reduction rate will increase the size of the training set. This mechanism can be used to generate input data in an implement of iterative learning algorithm.

3.1 Frequent Item Set Mining

The generation of frequent item sets from a transactional database is an important task of data mining and association rule detection. It filters the transactional database to remove the non-frequent items. Thus the generation of the frequent item sets is used to reduce the transactional database of search space. The reduced set of terminal symbols makes the rule detection algorithms faster and more efficient. On the other hand, the reduction also causes some lost of information. That is why the proper choice of the parameters of some algorithms has high importance and requires an expert user.

We think that there are some similar aspects in association rule detection and grammar inference. While the transactional database is a set of transactions of item sets. The corpus can be considered to be a set of sentences where a sentence is a sequence of the words. The difference is that in a transaction one item can occur only once. A word in a sentence can be repeated and the order of the words is given. These difference can influence the induction phase.

As our experimental results [10] already have showed the Inductive CYK algorithm [6] is much faster than the TBL [7, 8] and ITBL [5, 12] algorithms. The main reason of this difference is the iterative work. The algorithm allows the expansion of the learnt grammar later. The TBL, ITBL algorithms induce a grammar description from a closed training set and if the training set is modified, the entire learning process has to be repeated. Thus preprocessing of the training set could yield more efficient algorithms in grammar induction as well as in association rule detection.

3.1.1 Apriori Algorithm

The Appriori algorithm [1] is a well-known and simple frequent item set mining algorithm. It is based on the Apriori fact: "All subsets of a frequent set are frequent too". Its consequence is that a set cannot be frequent if it contains a non-frequent subset. On this basis, the Apriori algorithm uses a bottom-up approach to determine the frequent itemsets. It works iteratively and in each iteration determines a candidate set from the results of the previous step. The iteration has two steps, in the first it generates each n-length candidates from the $(n-1)$-length frequent itemsets. The generation of the candidate patterns is a costly process because $\frac{m*(m-1)}{2}$ $(n+1)$-length candidates can be generated from m n-length frequent itemsets. Experimental results show that there is a peak of the costs at $n = 2$. Then it uses exhaustive search to decide whether candidate is frequent or not.

3.1.2 FP-Growth

The FP-Growth algorithm [4] determines the frequent itemsets without candidate generation. This method is also based on the bottom-up approach because it grows the patterns from a conditional transactional database. It has two phases. In the first phase it filters and sorts the transactions, the rare items are removed and the transaction is represented as a string or sequence of items which begins with the most frequent item. Then it uses a recursive algorithm to determine the frequent itemsets. This method requires a conditional transaction database which contains transactions including the given items. Due to the recursive work, this algorithm requires substantial memory, but it is faster than the Apriori algorithm because candidate generation is not required.

3.2 Training Set Reduction Using Extended Regular Expressions

The previously detailed methods are tailored to frequent itemset mining, but not to text processing. While data mining processes structured data, the text mining processes unstructured documents. Grammar Induction yields the grammar description from positive and/or negative sentences which is given as sentences. Our assumption is that a method, which applies on text processing techniques, could be more efficient than the previous ones. Wildcard replacement is a recently proposed method to compress the corpus, thus it could be the base of a novel iterative algorithm. It is a internal phase in the Grammar Induction process and it yields a compressed corpus for the next steps.

Wildcard reduction is based on Regular Expressions. Regular Expressions is quite useful for string processing and are widely used in many contexts such as search by file name, `grep` command in UNIX systems or the `split` method of Java String object.The building blocks of Regular Expressions are simple structure elements, like blocks of symbols, negation, multiplicity or arbitrary symbols. Regular expressions can be considered as a compressed form for an arbitrary string, e.g. Hungarian number plates (`[A-Z]3-[0-9]3`) or telephone numbers ($\backslash+06-[0-9]2/[0-9]+$), etc. The regular expressions describe a set of sequences of symbols like formal grammars. There have been a great efforts to induce regular expressions from samples and it can be considered as regular language learning. Regular expressions are equivalent to the regular grammar class, thus this class is not powerful enough to model natural languages.

On the other hand, the regular expressions can be used as powerful tool in the discovery of frequent patterns. We can assume that the sentences of the training set are all of finite length. Thus each sentence can be converted into a valid regular expression containing only literal values. Having some regular expressions, a generalized expression can be formulated. For example, the source sentences

- The dog is running or the dog is sleeping.
- The dog is sleeping or the dog is running or the dog is barking.
- The dog is barking or the dog is sleeping or the dog is barking.
- ...

can be generalized into (the dog is [running, sleeping, barking]) (or the dog is [running, sleeping, barking])*. This generalization enables a more compact training set representation form. The regular expression formalism of the training sentences does not mean that the generated grammar is also a regular grammar. Using an extrapolation mechanism where the length of the training sets tends to infinite, the finite regular expression will be transformed into a higher order grammar. For example, having the series of regular expressions

$$A\{2\}BA\{2\}$$

$$A\{3\}BA\{3\}$$

$$A\{4\}BA\{4\}$$

$$A\{5\}BA\{5\}$$

$$A\{6\}BA\{6\}$$

the samples can be generalized into the expression

$$A\{n\}BA\{n\}$$

which is known CFG formula and not part of the regular language. As this formula is not a valid regular expression, we propose some extension of the base Regular Expression language. The proposed regular expression formalism is aimed at eliminating of irrelevant symbols and it supports an efficient generalization into CFG form.

Wildcard replacement method is based on the multiplicity operators of Regular Expressions and statistical approaches. There are four multiplicity operators in Regular Expressions: the one or zero (?), exactly one (no sign), one or more (+) and zero or more (*). Moreover, exact occurrence is also allowed. Based on these operators, the following three complex symbols are defined as:

word Matches exactly the same word once.
$? $ Matches exactly one arbitrary word.
$ * $ Matches two or more arbitrary words.

These novel symbols allow us to rewrite the training set into a more compact form. The frequent and the rare words are distinguished based on a parameter and the rare words will be replaced by a specific symbol. The frequent words are to remain unchanged. A rare word between two frequent ones is substituted by $?$ and a sequence of rare words is merged into the symbol $*$. Although Regular Expressions has the symbol `\w` which matches for any words and the pattern `\w{*}` represents arbitrary sequences of words, but `\w` does not distinguish the frequent and rare words. To extract patterns the distinction of frequent and rare words is necessary that is why $?$ and $*$ are introduced. Finally the $*$ sentences which contain only rare words are removed from the training set because they do not offer any information about the grammatical structure. The process yields a filtered training set where the sentences consists of at least one frequent word and zero or more symbols.

This compression has a statistical approach as well. This algorithm uses a parameter which is a percentage (has to be greater than zero and less than one) that affects the number of frequent words. It determines what percent of the sorted frequency table is frequent. When it is zero, the algorithm yields an empty training set because every word is rare. An increase in the value of this parameter increases the number of frequent words. Every grammar has statistical characteristics and this can be extracted easily from the sentences of the language.

Example This algorithm has been tested on two corpora. The first corpus contains SQL select statements (see Table 3). The second corpus is the English Bible which contains more than ten thousand sentences. We tested our method on these two training sets. The Table 3 shows the base and the filtered training set where the `select`, `from`, `car` words were frequent and the name of the fields are rare. The input parameter was 0.2 i.e. the words in the upper 20 % of the sorted frequency table were frequent.

Table 3 Wildcard reduction on SQL sentences

Sample	Replaced
	Select $?$ from car
Select * from car	Select $?$ from car
Select color from car	Select $?$ from car
Select price from car	Select $?$ from car
Select type from car	Select $?$ from car
Select brand from car	Select $?$ from car
Select price from car	Select $?$ from car
Select color price from car	Select $*$ from car

Table 4 Wildcard reduction measurements

Parameter (%)	Sentences	Words	Distinct words	Required time (ms)
Base	32290	791761	28334	No data
0	31102	184372	3	826
5	31102	184372	3	841
10	31102	184372	3	858
25	31102	268056	4	957
50	31102	322246	5	881
75	31102	322246	5	930
90	31102	515895	18	930
95	31102	614015	36	895
100	31102	790573	28333	901

The English Bible was used to measure the costs and the efficiency of the filtering algorithm with different parameters. It is easy to see that it is a linear algorithm because it requires two loops. In the first loop the frequency table of the words and the set of frequent words are determined. The second loop yields the filtered grammar by at replacement of the rare words. Thus the execution time of the algorithm is $O(2n) = O(n)$ where n denotes the size of the training set, so it is a linear algorithm. The compression rate is a key parameter in our measurements because the number of resulting sentences and of distinct words has a significant effect on the learning process. Our expectation was that this algorithm would reduce the number of vocabulary items as well as the lengths of the sentences. Filtered training set would converge to the base training set when the parameter converged to 1 and the required time would be constant because the corpus is fixed.

The results of the measurements are given in Table 4 and it confirms our hypothesis. The column Sentences denotes the number of the sentences in the reduced training set. The column Words denotes the cumulative number of the words and increases with the count of the distinct words. The measurements showed that there is a sharp gap in the number of words and distinct tokens which in this case is between 95 and 100 %. This can be explained by Zipf's law which says that there are a few short and frequent words, and there are many rare word in natural languages. There are only a few words in the corpus which are used frequently and the largest part of the vocabulary is the set of the rare words. The algorithm required almost the same time with each parameter, so the runtime does not depend on this value.

4 Pattern Distillation Based on Pattern Intersection

The incoming training sentence is a sequence of terminal symbols. At this level, both the terminal symbols and the sentences can be considered to be patterns. The goal of pattern distillation is to determine the intermediate patterns as sequences of other patterns. A pattern can be considered to be a cluster of other patterns that are used to compose the grammar structure of the language. A main characteristic of a pattern is its relatively high frequency. If a sequence is rare, no pattern is created to cover that sequence, as it will not result in significant grammar compression. Another consideration is the existence of a hierarchical structure between the patterns based on the containment or subset relationship. The particular hierarchical structures can be merged into a lattice of pattern elements.

In the proposed grammar induction system, the frequent subsequences will be generated with a special intersection operation. The intersection will yield new subsequences related to the two patterns. The result set consists of two kinds of patterns. The first group contains the pattern found in both operands. The second group includes patterns found in only one operand. In this group, there is a substitution relationship defined among the elements. This is a bipartite relationship; two patterns are in relation if they are located in the same relative positions. Based on the substitution relationship, a substitution graph can be constructed for the training set. Each edge of the graph is assigned a weight value, where the weight value corresponds to the frequency of the given pair. The nodes of the graph refer to the patterns. Having the weighted substitution graph, the patterns can clustered into groups. A new cluster can be considered to be a new pattern which can substitute its members. After replacing the cluster's members with the cluster element in the graph, clustering can be continued at the new level. Thus the resulting structure will be a hierarchical clustering of the patterns.

Regarding the intersection operation, the operands are sequences and not sets. Thus, the elements have positions and the same element may be repeated in the sequence. The intersection operation to define the maximum common set of subsequences may have several interpretations. The paired subsequences must have the same position number in both operands. For example, for the operand sequences

$$s_1 = AAABBAABB, s_2 = AABB, \quad (1)$$

there are many valid assignment mappings. Some possible common subset assignments are as follows:

$$s_1 = (AA)ABBAA(BB); s_2 = (AA)(BB) \quad (2)$$

$$s_1 = A(AABB)AABB; s_2 = (AABB) \tag{3}$$

$$s_1 = AAABB(AABB); s_2 = (AABB) \tag{4}$$

In our solution, the following approaches were used to reduce the ambiguity. First, every repetition is symbolized with the symbol A* symbol which differs from symbol A. Based on this notation, the previous sample can be given as.

$$s_1 = A*B*A*B*; s_2 = A*B* \tag{5}$$

which can have only three valid assignments:

$$s_1 = (A*B*)A*B*; s_2 = (A*B*) \tag{6}$$

$$s_1 = A*B*(A*B*); s_2 = (A*B*) \tag{7}$$

$$s_1 = (A*)B*A*(B*); s_2 = (A*)(B*) \tag{8}$$

The second consideration is that the intersection is implemented in a recursive way. At a given level, the first occurrences of the longest matching subsequence is determined. The single level extended intersection can be given as a pair of

$$s_1 \bigwedge s_2 = ((s_1 \bigcap s_2), (s_1 \Delta s_2)) \tag{9}$$

where

$$\begin{aligned} \bar{s} = s_1 \bigcap s_2 : \bar{s} \subset s_1, \bar{s} \subset s_2, \neg\exists s' : (|s'| > |\bar{s}|, s' \subset s_1, s' \subset s_2), \\ \forall s''(s'' \subset s_1, s'' \subset s_2, |s''| = |\bar{s}|) : pos(s'') > pos(\bar{s}) \end{aligned} \tag{10}$$

The single level intersection results in a common part and four distinct parts in general. The not distinct parts can be considered to be a tupple of set differences:

$$\bar{s} = s_1 \Delta s_2 : \bar{s} = \{s_{1,1\ldots k_1}, s_{2,1\ldots k_2}, s_{1,v_1+1\ldots n_1}, s_{2,v_2+1\ldots n_2}\} \tag{11}$$

where the common substring is located between the positions (p_1, v_1) in s_1 and between (p_2, v_2) in s_2. The symbols k_1, k_2 denote the length of the operand sequences. In the multilevel intersection, the corresponding left and right remaining parts are processed in a recursive way. If for a level i, the related intersection is denoted by

$$s_{1,i} \wedge s_{2,i} = ((s_{1,i} \cap s_{2,i}), (s_{1,i} \Delta s_{2,i})) \tag{12}$$

then for the sub-level *(i+1)*

$$s_{1,i+1} = s_{1,1\ldots k_1,i}; s_{2,i+1} = s_{2,1\ldots k_2,i} \tag{13}$$

or

$$s_{1,i+1} = s_{1,v_1+1...n_1};\ s_{2,i+1} = s_{2,v_2+1...n_2,i} \tag{14}$$

The empty substring will be denoted by ϵ. As at decomposition structure can be described with a hierarchy, there are more decompositions at the higher level. The index i* denotes the union of all components at the level i. Having two sequences, the multi-level extended intersection is equal to the union of generated common parts and the disjoint parts at the highest level:

$$s_1 \wedge^* s_2 = \cup_i((s_{1,i^*} \cap s_{2,i^*}), (s_{1,i\,\max^*} \Delta s_{2,i\,\max^*})) \tag{15}$$

Due to efficiency considerations, usually only a limited decomposition high is processed. For example, with imax = 2, the sentences

$$\begin{array}{r}\text{'SELECT a, b FROM e WHERE c'}\\ \text{'SELECT a FROM d',}\end{array} \tag{16}$$

the extended intersection for level i=1 is equal to

$$\begin{array}{r}s_{1,1} \cap s_{2,1} = \{\text{'SELECT a'}\}\\ s_{1,1} \Delta s_{2,1} = \{\epsilon, \epsilon, \text{',b FROM e WHERE c', 'FROM d'}\} =\\ \{\text{', b FROM e WHERE c', 'FROM d'}\}\end{array} \tag{17}$$

For level i = 2, we get

$$\begin{array}{r}s_{1,2} \cap s_{2,2} = \{\text{'FROM'}\}\\ s_{1,2} \Delta s_{2,2} = \{\epsilon, 'b', 'd', \text{'e WHERE c'}\} = \{\text{'b', 'd', 'e WHERE c'}\}\end{array} \tag{18}$$

Thus the result is the following

$$\begin{array}{r}s_1 \cap^* s_2 = \{\text{'SELECT a', 'FROM'}\}\\ S_1 \Delta^* s_2 = \{\text{'b', 'd', 'e WHERE c'}\}\end{array} \tag{19}$$

The corresponding lattice is presented in Fig. 1.

The substitution relationship is defined between two non-common parts from different operands, if they belong to the same position. In the example, this relationship can be discovered only between 'e WHERE c' and 'd'. This relationship is defined with a dashed line in the lattice (Fig. 2).

Once we have the complex containment lattice and substitution graph, the next step is to discover the pattern clusters in order to introduce new generalized patterns. The data source for clustering is the substitution graph with weighted edges. The simplest way of clustering is the application of the HAC method [11]. The HAC method works in the following steps:

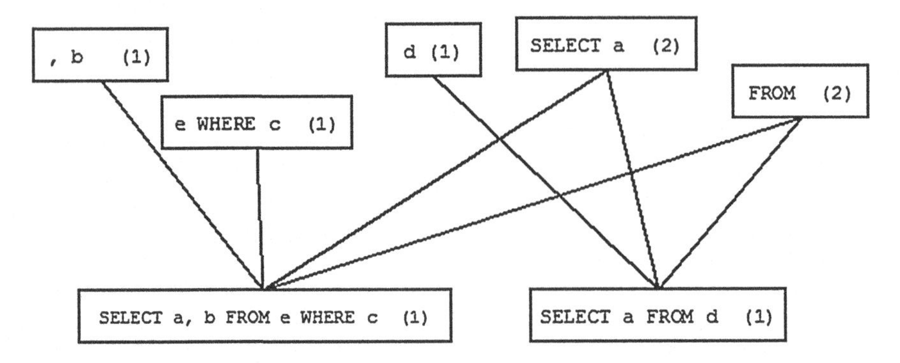

Fig. 1 The containment lattice

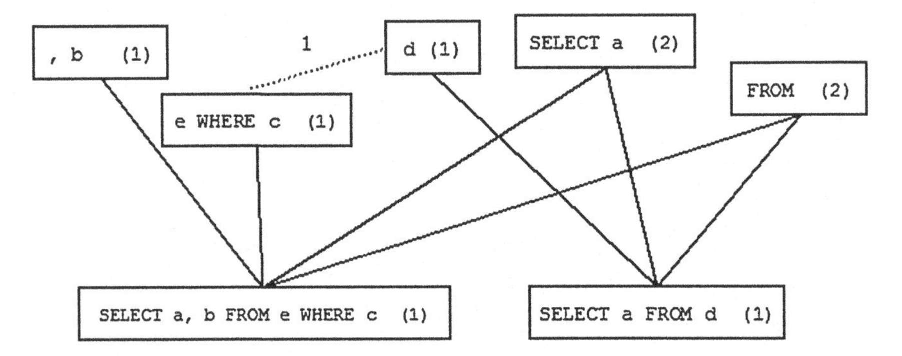

Fig. 2 The containment lattice and substitution graph

- initially every node is a separate cluster
- join the clusters with the nearest distance from each other into a single new cluster
- repeat the merge operation until a threshold value is met.

In the case of the substitution graph, the distance between two nodes is calculated from the weight value of the connecting edge. The highest weight value corresponds to the smallest distance. In our HAC module, a minimum value for the weight is the termination condition.

The generated lattice elements can be used as frequent input patterns in the grammar induction algorithm. The frequency attribute of the lattice nodes can be used to eliminate infrequent patterns from the training set. Having the frequent patterns, the grammar induction module should discover the structural relationship between the patterns. The containment relationship is the base parameter in building up the grammar tree. As the lattice structure conveys also this kind of information, the lattice can be considered as specific type of grammar structure.

Regarding, the execution cost complexity of the proposed method, the first component is the analysis of the single level extended intersection. The problem of finding the longest common subsequence relates to the longest common substring problem.

The usual algorithm [3] for this problem is to generate first a prefix tree to store both operand strings. The building of the prefix tree requires $O(M)$ cost where M denotes the length of the operand strings. In the next phase, the longest common substring part is located with the application of a dynamic programming solution. In the corresponding table for dynamic programming, the rows and columns represent the different character positions within the string. This step requires $O(M_1 M_2)$ cost.

The next phase is to repeat this segmentation at different levels. Thus, the extended intersection of two sequences takes $O(LM_1 M_2)$ elementary steps, where L denotes the iteration threshold value. Having N input training sentences and W terminals symbol words, the generated lattice may contain in the worst case $O(2^W)$ nodes. Thus the construction of the whole lattice may require in the worst case $O(2^W LM_1 M_2)$ steps.

In order to achieve a tractable solution, additional cost reduction methods are required. In our prototype system, the following reduction methods were applied:

- setting the maximum level of extended intersection to a low value
- removing the elements below a given frequency threshold
- setting the maximum distance to a high level; the resulting clusters will be used as symbols in the grammar tree.

The test experience shows that the current version of the algorithm is suitable only for small size problems because of the high complexity of the applied lattice structure.

5 Conclusion

In this chapter two approaches are proposed to extract the patterns from the training set in grammar induction systems. The frequent itemset processing systems and the regular expressions give the background for the first method. Both the grammar induction and the generation of association rules share a common aspect because they yield a rule set. The determination of frequent itemsets is the preprocessing step of the association rule generation process. The Apriori and the FP-Growth may be possible solutions but these methods are costly and work on sets instead of sequences. A novel method was proposed and tested based on the multiplicity operators of regular expressions to filter the training set. The experimental results show that it is a fast algorithm. However, the proper choice of its parameter requires further investigation.

The other method is based on an approach of concept lattices. This method extracts the common and the unique parts of the patterns or sequences into a new node in the lattices. The common parts are the bases of the rules and the unique parts can be changed. This algorithm has been presented with an example.

Acknowledgments The described work was carried out as part of the TÁMOP-4.2.2/B-10/1-2010-0008 project in the framework of the New Hungarian Development Plan. The realization of this project is supported by the European Union, co-financed by the European Social Fund.

References

1. Agrawal, R., Srikant, R., et al.: Fast algorithms for mining association rules. In: Proceedings of the 20th International Conference on Very Large Data Bases, VLDB, vol. 1215, pp. 487–499 (1994)
2. D'Ulizia, A., Ferri, F., Grifoni, P.: A survey of grammatical inference methods for natural language learning. Artif. Intell. Rev. **36**, 1–27 (2011)
3. Gusfield, D.: Algorithms on strings, trees and sequences: computer science and computational biology. Cambridge University Press, Cambridge (1997)
4. Han, J., Pei, J., Yin, Y.: Mining frequent patterns without candidate generation. ACM SIGMOD Rec. **29**, 1–12 (2000)
5. Jaworski, M., Unold, O.: Improved tbl algorithm for learning context-free grammar. In: Proceedings of the International Multiconference on Computer Science and Information Technology, ISSN 1896–7094, pp. 267–274 (2007)
6. Nakamura, K., Matsumoto, M.: Incremental learning of context free grammars based on bottom-up parsing and search. Pattern Recogn. **38**, 1384–1392 (2005)
7. Sakakibara, Y., Kondo, M.: Ga-based learning of context-free grammars using tabular representations. In: Machine Learning-International Workshop then Conference, pp. 354–360, Morgan Kaufmann Publishers (1999)
8. Sakakibara, Y.: Learning context-free grammars using tabular representations. Pattern Recogn. **38**, 1372–1383 (2005)
9. Solan, Z., Horn, D., Ruppin, E., Edelman, S.: Unsupervised learning of natural languages. Proc. Nat. Acad. Sci. U.S.A. **102**(33), 11629–11634 (2005)
10. Tóth, Zs., Kovács, L.: CFG extension for META framework. In: IEEE 16th International Conference on Intelligent Engineering Systems, pp. 495–500 (2012)
11. Tsoumakas, G., Katakis, I., Vlahavas, I.: Mining multi-label data. In: Rokach, L., Maimon, O. (eds.) Data Mining and Knowledge Discovery Handbook, pp. 667–685. Springer, Berlin (2010)
12. Unold, O., Jaworski, M.: Learning context-free grammar using improved tabular representation. Appl. Soft. Comput. (2009)

Advanced 2D Rasterization on Modern CPUs

Péter Mileff and Judit Dudra

Abstract The graphics processing unit (GPU) has become part of our everyday life through desktop computers and portable devices (tablets, mobile phones, etc.). Because of the dedicated hardware visualization has been significantly accelerated and today's software uses only the GPU for rasterization. Besides the graphical devices, the central processing unit (CPU) has also made remarkable progress. Multi-core architectures and new instruction sets have appeared. This chapter aims to investigate how effectively multi-core architecture can be applied in the two-dimensional rasterization process and what the benefits and bottlenecks of this rendering model are. We answer the question of whether it would be possible to design a software rendering engine to meet the requirements of today's computer games.

1 Introduction

Computer graphics has undergone dramatic improvements over the past few decades, and one important milestone was the appearance of graphic processors. The main objective of the transformation was to improve graphical computations and visual quality. Initially, the development process of the central unit was far the fast-paced evolution of today. So based on industry demands, there was a need for dedicated hardware to take over the rasterization task from the CPU.

Graphical computations have requirements different from the other parts of the software. This allowed graphics hardware to evolve independently from the central unit, opening new opportunities to developers, engineers and computer game

---

P. Mileff (✉)
Department of Information Technology, University of Miskolc,
Miskolc-Egyetemváros, Miskolc 3515, Hungary
e-mail: mileff@iit.uni-miskolc.hu

J. Dudra
Department of Structural Integrity, Bay Zoltán Non-profit Ltd,
Iglói út 2, Miskolc 3519, Hungary

G. Bognár and T. Tóth (eds.), *Applied Information Science, Engineering and Technology*, Topics in Intelligent Engineering and Informatics 7, DOI: 10.1007/978-3-319-01919-2_5,

designers. From the perspective of manufacturers and the industry, primarily speed came to the fore against programming flexibility and robustness. So in recent years the development of video card technology focused primarily on improving the programmability of its fixed-function pipeline. As a result, today's GPUs have a quite effectively programmable pipeline supporting the use of high-level shader languages (GLSL, HLSL and CG).

Today technological evolution proceeds in quite a new direction, introducing a new generation of graphics processors, the general-purpose graphics processors (GPGPU). These units are no longer suitable only for speeding up the rendering, but have capabilities for general calculations similar to those of the CPU. Simultaneously with the soaring of the graphical chips, central units also have evolved. Although this line of development was not as spectacular as the development of GPUs, it was undoubtedly important. The appearance of a new GPU or video card was always surrounded by major media advertising compared to the attention given to central units. For CPUs, initially two development lines have evolved. The first approach considered it appropriate to increase the number of central processing units (Multiprocessing Systems) and then the objective was to increase the cores inside the central unit (Multicore processors). The first dual-core processors have appeared starting in 2005. A multicore system is a single-processor CPU that contains two or more cores, with each core housing independent microprocessors. A multicore microprocessor performs multiprocessing in a single physical package. Multicore systems share computing resources that are often duplicated in multiprocessor systems, such as the L2 cache and front-side bus. Multicore systems provide performance that is similar to that of multiprocessor systems but often at a significantly lower cost, because a motherboard with support for multiple processors, such as multiple processor sockets, is not required.

All of these gave a completely new direction to software development, making the evolution of multi-threaded technology possible. Today we can say that the core increasing process has significantly contributed to improving the user experience of the operating systems (multitasking) and the development of multi-threaded applications. Multi-threaded software which exploits properly the hardware features can far outperform the performance of software applying the classical, one-thread approach. Observing this trend we can see that processor manufacturers and design companies have established themselves in the production and design of (multi) core based central units. Today's mobile devices also have multiple cores (2–4) and in case of PCs the number of cores can reach eight units due to the Hyper-Threading technology.

Besides the core increasing tendency, the processor manufacturers responded with extended instruction sets to market demands, making faster and mainly vectorized (SIMD) processing possible also for central units. Almost every manufacturer has developed its own extension, like the MMX and SSE instruction families, which were developed by Intel and are supported by nearly every CPU. Initially, AMD tried to boost its position with its 3DNow instruction set, but nowadays the direction of development of mobile central units is the Vector Floating Point (VFP) technology and the SSE such as the NEON instruction set initially introduced in ARM Cortex-A8 architecture. For PCs Advanced Vector Extensions (AVX) open up again great new opportunities in extending performance.

Due to new technologies, software can reach multiple speedups by properly exploiting the hardware instruction set. It is therefore appropriate to examine the speedup possibility of the rasterization process. Is there any point in developing a software renderer capable of meeting the needs of today's computer games? This chapter investigates the practical implementation issues of two-dimensional rasterization. A special optimization solution is presented, which helps to improve non-GPU-based rendering for transparent textures, heavily utilizing the CPU cores for higher performance.

2 Related Works

Software-based image synthesis has existed since the first computers and was focused even more with the appearance of personal computers, up until about 2003. After this time almost all the rendering techniques became GPU based. However, there were many interesting software renderers created during the early years. The most significant results were the Quake I and Quake II renderers in 1996 and 1998, which are the first real three-dimensional engines [8]. The rendering system of the engines was brilliant compared to the computer technology of that day, and was developed under the coordination of Michael Abrash. The engine was typically optimized for the Pentium processor family, taking advantage of the great MMX instruction set. The next milestone in computer visualization was the Unreal Engine in 1998 with its very rich functionality (colored lighting, shadowing, volumetric lighting, fog, pixel-accurate culling, etc.) [12]. Today Unreal technology is a leader in the area of computer graphics.

After the continuous headway made in GPU rendering, software rasterization was increasingly losing ground. Fortunately, there are some notable great results today as well, such as Swiftshader by TrasGaming [2] and the Pixomatic 1, 2, and 3 renderers [14] by Rad Game Tools. Both products are very complex and highly optimized, utilizing the modern threading capabilities of today's Multicore CPUs. The products have dynamically self-modifying pixel pipelines, which maximizes rendering performance by modifying its own code during runtime. In addition, Pixomatic 3 and Swiftshader are 100 % DirectX 9 compatible. Unfortunately, since these products are all proprietary, the details of their architectures are not released to the general public.

Microsoft supported the spread of GPU technologies by the development of DirectX, but in addition, its own software rasterizer (WARP) has also been implemented. Its renderer scales very well to multiple threads and it is even able to outperform low-end integrated graphics cards in some cases [3].

In 2008 based on problem and demand investigations, Intel aimed to develop its own software solution based video card within the *Larrabee* project [5]. In a technological sense, the card was a hybrid between the multi-core CPUs and GPUs.

The objective was to develop an x86 core (many) based fully programmable pipeline with 16 byte wide SIMD vector units. The new architecture made it possible for graphic calculations to be programmed in a more flexible way than GPUs with an x86 instruction set [4].

Today, based on the GPGPU technology, a whole new direction is possible in software rendering. Loop and Eisenacher [17] describe a GPU software renderer for parametric patches. The FreePipe Software rasterizer [10] focuses on multi-fragment effects, where each thread processes one input triangle, determines its pixel coverage, and performs shading and blending sequentially for each pixel. Interestingly, recent work has also been done by NVidia to create a software pipeline which runs entirely on the GPU using the CUDA software platform [7]. The algorithm uses the popular tile-based rendering method for dispatching the rendering tasks to the GPU. Like any software solution, this allows additional flexibility at the cost of speed.

A new SPU (Cell Synergistic Processor Unit) based deferred rendering process has been introduced in today's leading computer game, Battlefield 3 [13]. Its graphical engine, a Frostbite 2 engine, makes it possible to handle a large number of light sources effectively and optimized. In [1] a modern, multi-thread tile based software rendering technique is outlined where only the CPU is used for calculations and had great performance results.

Thus, recent findings clearly support the fact that CPU-based approaches are ready to come back in order to improve performance and flexibility. So, the aim of this chapter is to investigate performance in the area of 2D software rendering utilizing today's CPUs.

3 Basics of 2D Rendering

The name 'software rasterization' originates from the imaging process where the entire image rasterization process, the whole pipeline, is carried out by the CPU instead of target hardware (e.g. a GPU unit). In this case the graphics card is responsible only for displaying the generated and finished image based on a framebuffer array located in the main memory. The main memory holds also the shape assembling geometric primitives in the form of arrays, structures and other data, ideally in an ordered form. The logic of image synthesis is very simple: the central unit performs the required operations (coloring, texture mapping, color channel contention, rotating, stretching, translating, etc.) on data stored in the main memory, then the result is stored in the *framebuffer* (holding pixel data) and the completed image is sent to the video controller.

A framebuffer is an area in the memory which is streamed by the display hardware directly to the output device. So its data storage logic needs to meet the requirements (e.g. RGBA) of the formats supported by the video card. To send the custom framebuffer to the video card, several solutions have arisen in practice. First, we can use the operating system routines (e.g. Windows—GDI, Linux—Xlib) for transfer, but it is strongly platform dependent. This method requires writing the bottom layer of the software separately for all the operating systems. A more elegant solution is to use OpenGL's platform-independent (e.g. GLDrawPixels or Texture) [6] or DirectX (e.g. DirectDraw surface or Texture) solutions.

3.1 Benefits of Software Rendering

Although software rendering is rarely applied in practice, it has many advantages over the GPU-based technology. The first and most emphasized point is that there is less need to worry about compatibility issues, because the pipeline stages are processed entirely by the CPU. In contrast with the GPU, the CPU's structure changes less rapidly, thus the need for adapting to any special hardware/instruction set (e.g. MMX, SSE, AVX, etc.) is much lower. In addition, these architectures are open and well-documented, unlike the GPU technology.

The second major argument is that image synthesis can be programmed uniformly using the same language as the application, so there is no restriction on the data (e.g. maximum texture size) and the processes compared to GPU language shader solutions. Every part of the entire graphics pipeline can be programmed individually. Because displaying always goes through the operating system controller, preparing the software for several platforms causes fewer problems. Today's two leading GPU manufacturers publish their drivers only in closed form, which leads to significant problems in performance with the Linux platforms. Driver installation is not easy on certain distributions; the end of the process is often the crash of the entire X server. The alternatively available open source drivers are limited in performance and other areas.

In summary, software rendering allows more flexible programmability for image synthesis than GPU technology.

3.2 Disadvantages of Software Rasterization

The main disadvantage of software visualization is that all data are stored in the main memory. Therefore in case of any changes of data the CPU needs to contact this memory. These requests are limited mostly by the access time of the specific memory type. When the CPU needs to modify these segmented data frequently, this can cause a significant loss of speed.

The second major problem, which originates also from the bus (PCIe) bandwidth, is the movement of large amounts of datasets between the main and the video memory. Within the period of one second the screen should be redrawn at least 50–60 times, which results in a significant amount of dataflow between the two memories. In the case of a 1024×768 screen resolution with 32 bit color depth, one screen buffer holds 3 MB of data.

Moreover, developing a fast software rendering engine requires lower level programming languages (e.g. C, C++, D) and higher programming skills. Because of the techniques used, it is necessary to use operating system-specific knowledge and coding.

4 Overview of 2D Rendering

Two-dimensional visualization plays an important role in addition to today's modern three-dimensional rasterization. Computer applications using graphical menu or windowing systems belong in this area, but the most obvious example is the desktop of the operating system. Undoubtedly the main users of the technique are the two-dimensional computer games. Over the years, there has been an increasing demand to improve the visual quality of the virtual world. Today a complex game operates on a large number of continuously changing and moving sets of objects. Thus the rasterization process consumes significant system resources, changing dynamically depending on the moving objects. Typical features of the renderer of these complex systems are a high screen resolution, large texture sets, animations and transformations to achieve better user experience. The screen resolution has been also increased. While in the past the 320×200 and 640×480 dimensions were sufficient, nowadays large high-quality textures (32 bit) are indispensable for higher screen resolutions (e.g. higher than 1024×768). All of these increase the requirements for performance, inducing continuous development of the rasterization models and techniques.

Based on the needs of computer games, in the following it will be shown what the main difficulties of two-dimensional rendering are and why the rasterization stage is so performance intense.

4.1 Characteristics of 2D Rendering

Two-dimensional rendering operates on images (textures) and objects (animations) using 2D algorithms. During the image generation process, the graphics engine is responsible for objects being drawn into the framebuffer one by one, based on a predefined drawing logic. So the final image is created as a combination of these. In all cases, textures are stored in the main memory, represented as a block of arrays. Arrays contain color information about the objects; their size depends on the quality of the texture. Today software works with 32-bit (4 bytes—RGBA) type color images, where image resolution can be up to 1024×768 pixels depending on the requirements of the items to be displayed.

Based on color channel information, textures can be divided into two types: images containing some transparent area (e.g. cloud, ladder), and images without transparent areas. The distinction is important because the rasterization and optimization methods differ for these types. In the following the implementation logic will be shown.

4.2 Rasterization Model of Non-opaque Textures

Textures without transparent areas use only RGB color components, or the alpha value of all the pixels is maximal. So the image does not contain any transparent pixels ('holes'). This information is very important because it fundamentally determines the display logic. This means that any two objects can be drawn on each other without merging any colored pixels of the overlapping objects. The rendering process will be significantly faster and simpler.

The rasterization of images without transparent areas is relatively simple; the entire texture can be handled at once in one or more blocks and not pixel by pixel. The memory array of the texture is moved into the frame buffer using memory copy operations (e.g. C—*memcpy()*).

There exists only one criterion: to avoid framebuffer over addressing, attention should be paid to the edges of the screen and during copying, the data should be segmented based on the object's position. If any object is located out of the screen bounding rectangle in any direction, a viewport culling should be performed row by row at the texture. Although this requires further calculations, the solution is still fast enough. So this method was preferred in the early computer games.

4.3 Rasterization Model of Textures Having Transparent Areas

In case of transparent textures, the part of the image contains transparent point groups, or the pixels of the image are opaque in some intensity. To implement the first category, a pre-selected but unused color has to be applied to mark the transparent pixels, this is called colorkey. Today, this is achieved using the alpha (A) channel associated with the image.

The role of this type of textures has increased today: they are used in many areas in order to improve the visualization experience (e.g. window shadows, animations with blurred edges, and semi-transparent components). Handling this extra information is not more complicated but is more computing intensive. The reason is that transparent and non-transparent areas can arbitrarily vary within a texture image (character animation, particle effects, etc.). Due to this, the rendering process is made at per-pixel level because transparent or semi-transparent parts of the objects should be merged with the overlapped pixels. Figure 1 illustrates the problem.

The graphical engine assesses the graphical objects pixel by pixel and generates the final image. The disadvantage of this is that many elements, which can also consist of many points, have to be drawn on the screen. The per-pixel drawing requires thousands of redundant computations and function calls. For each pixel the color information should be read from memory, then depending on the environmental data its position should be determined and finally the color should be written into the framebuffer (e.g. pFrameBuffer[y * screenWidth + x] = color).

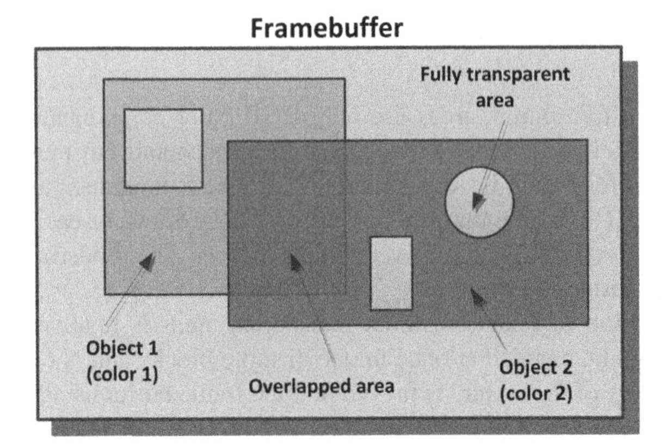

Fig. 1 Overlapping RGBA textures

For this reason, this technique does not possess high enough performance to meet the requirements of today's complex computer games, where up to 100 different moving objects should be drawn simultaneously on the screen. Too many small operations should be performed, which consumes CPU resources.

5 Accelerate 2D Visualization Applying Thread Management

Given the process and the difficulties of rasterization, the question arises whether it is possible to find a more effective solution which is able to meet the growing demands of today's computer games. As long as the above model is implemented using classical programming technology, performance results will certainly not be satisfactory because the model does not take into account the characteristics of available hardware.

The following simple investigation will confirms this fact: while writing the chapter, we implemented a simple one-threaded rasterizer (classical model). The objective was to make performance profiling utilizing transparent textures. The results showed clearly that a software running time of 97 % exposes the rendering of the transparent textures: calculating color and position information of the pixel and writing them onto the framebuffer. This by itself does not indicate the inefficient use of the technology. Therefore, as a further investigation, it was observed how many CPU cores were active and to what extent during the runtime. The results showed that at the same time only one CPU core load was high, the load of the others was minimal. This confirms the fact that the simple rasterization model cannot properly take advantage of the features of the hardware.

In the field of parallelization, central units have developed substantially during the past years. The number of cores is continuously increasing and the instruction sets are slowly but also continuously evolving, e.g. the AVX instruction set applyied at Intel Sandy Bridge and AMD Bulldozer processors. Each core has its own cache memory. All of these provide a good basis for establishing intensified parallel computations. Today, having four cores in a central unit of a desktop computer is natural and mobile devices are slowly catching up in this area. An example is the Tegra 4, offering quad-core Cortex A15. It is therefore clear that these should be used in the rasterization process. It is indispensable to design and implement the rasterization model in such a way so as to be able to take into account the opportunities of today's hardware parallelism and dynamically adapt to them.

A software renderer developed by applying parallel technology properly is expected to achieve significantly better results than the classical approach. In the following such a model is outlined, supported by detailed test results.

5.1 The Model of the Distributed Rasterizer

In the course of rasterization it is appropriate to develop a distributed model for the logic of the rendering engine. A solution should be designed which can be built from well parallelized processes. Naturally, the degree of parallelization has a theoretical maximum upper limit defined by Amdahl's Law [15]: the speedup of a program using multiple processors in parallel computing is limited by the time needed for the sequential fraction of the program. Therefore it is important to investigate first to what extent the rendering process corresponds to the principles of parallelization.

The technology of parallelization is based on the appropriate utilization of CPU cores. This can be achieved most effectively by using hardware threads. However, thread management raises several optimization issues. As an important rule, the core rasterization model should be designed to take into account the potential number of cores in the particular hardware central unit. The parallelization is not optimal when the working threads of an application exceed the number of CPU cores (in the case of Intel processors this includes the virtual cores given the Hyper-Threading technology). When the number of logical threads reaches the number of available hardware threads, the performance slowly starts to decrease because of context switches [15].

Fortunately the process of two-dimensional visualization is simple from the mathematical point of view, therefore parallelization can be adapted more easily than in the case of 3D. The rendering is based on writing pixels into the framebuffer. Since pixels are independent of each other, thus the process of the rasterization can be well parallelized under certain criteria. The most important criterion required is that the synchronization between threads should be minimal. To achieve this, the only requirement is that the same elements of the framebuffer cannot be written by threads at the same time. If we can solve the problem that the threads should not wait for each other during the pixel writing process, the performance benefit is expected to be significant. The result is a multi-threaded rendering model implementing distributed rasterization.

Because the rendering process is mainly slowed by the rasterization of transparent objects, hereafter the chapter primarily deals with this category.

5.1.1 Advanced Distributed Rendering Model

To design a distributed rasterization model achieving minimal synchronization, we should start from the logical division of the visualization area similar to GPU hardware and Tile Based rendering. Because this division also determines the logical division of the framebuffer, the areas should be designed to be independent. In this case, if the rasterization of the area is performed by separate processing threads, the necessary synchronization between threads will be minimal. The reason for this is that none of the render threads will do pixel operations on areas belonging to other threads.

Starting from the former idea that today a 4-core central processing unit is standard, we consider four identical-sized areas for the screen division. Thus the rasterization process can be performed using four threads. Naturally, the efficiency of this process is optimal when the number of logical screen areas corresponds to the CPU cores. Figure 2 shows the logical steps of the rasterization process.

The first part of the applied graphics pipeline is identical to the classical solution. The position and orientation calculation of the 2D objects is performed and the objects are located in the region of the screen. If at least one pixel is visible of the object's texture, its rasterization is indispensable. However, this process differs

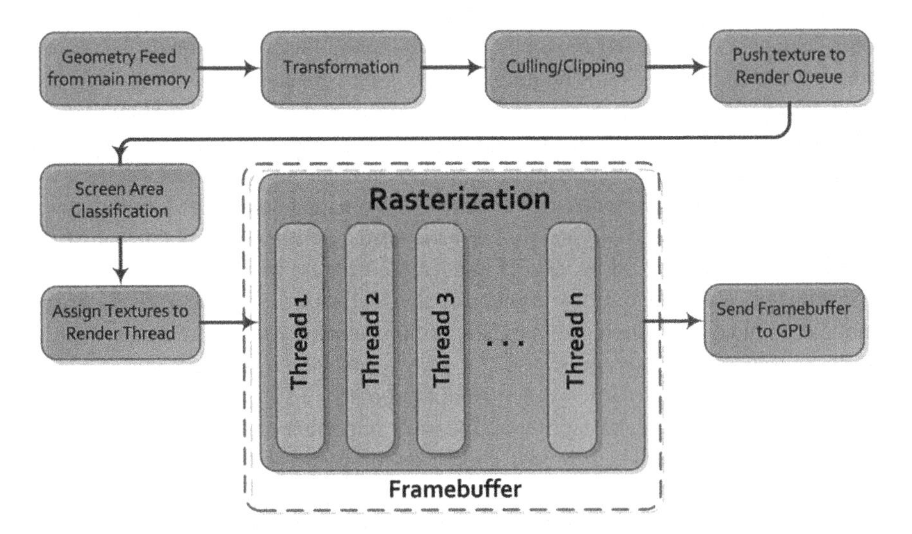

Fig. 2 Distributed rasterization logic

from the classical method. While previously calling a *Draw()* method immediately writes the texture of the object into the framebuffer, this solution requires a container that collects and holds these textures on a list during the rasterization. The actual rasterization occurs when all of the object's texture is on the list intended for drawing.

The basis of the parallel rasterization is that the different areas of the framebuffer can be written independently and parallel with each other. To accomplish this, a fast classification algorithm is required, which determines the area where the object belongs and its render thread. However, the classification raises a further question. Surely there will be objects whose images overlap the logical areas. So during the classification process the texture will be associated with both threads for rendering, which violates the rules of parallelism laid down. One solution to these problems can be a simple model where the classification algorithm associates the textures with the areas in such a way that the elements belonging to more than one area are associated with the main thread of the application.

Textures belonging to this thread should be drawn only before or after the rendering process of the other threads. Because of overlapping, the writing process of the thread affects the other thread's logical framebuffer parts. Although the solution is operational and fast, it is not applicable in every case. In computer games, a predefined rendering sequence is usually required (e.g. an airplane flies in front of certain clouds and behind certain clouds). This solution cannot keep the rendering order because of the overlapped areas. To extend the model in order to support the rendering order would require too much communication between threads, killing parallel performance.

In order to keep the advantages of the parallelism, the overlapping problem should be solved. Since we are in the two-dimensional screen space, the texture of each object can be described with a rectangle. This is the area bounded by the texture's width and height.

An effective approach to the problem is to cut the image of the objects along the logical area borders. For this, it is necessary to modify the classification algorithm. While previously all the textures overlapping the logical areas were associated with the main render thread, this model associates these textures with any logical area that is overlapped by the image. The most complex case is when the texture overlaps all four areas of the screen. This requires cutting the texture into four different parts and associating them with four threads. Figure 3 shows the logic of texture cutting and thread association.

During the visualisation process further checks and calculations are required. Because overlapping textures can belong to a logical area, during pixel level rasterization it is necessary to determine the exact pixel borders of the rendering. This avoids conflict with other threads. All of these operations require extra performance from the CPU compared to the previous simple solution, but this does not radically affect the rasterization speed. However, it is clearly visible that this model keeps the rendering order. If the display list is arranged in the correct order before the classification process, the order remains unchanged during rasterization.

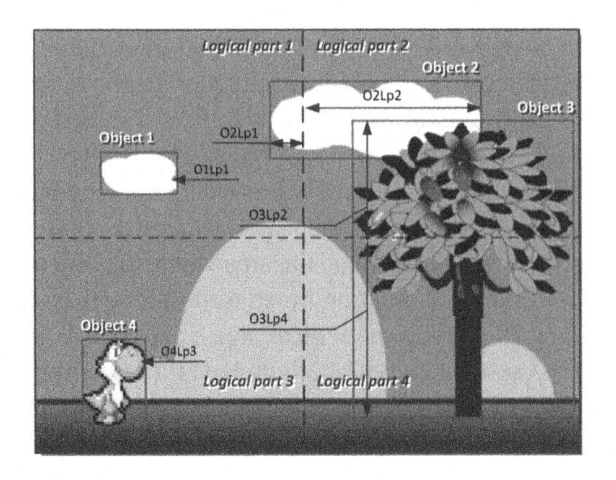

Thread 1	Thread 2	Thread 3	Thread 4
O1Lp1	O2Lp2	O4Lp3	O3Lp4
...	...	...	...
...	...	...	...
...	...	...	...
...	...	...	...
O2Lp1	O3Lp2	...	...
...	...	...	...
...	...	...	...
...	...	...	...
...	...	...	...

Fig. 3 Rasterization logic

5.1.2 Properties of the Rendering System

Naturally, since threads are used in the visualization system, synchronization between threads is inevitable. The model requires two synchronization points. First, when the threads are assigned their tasks after the classification process and receive the 'start processing' signal. The second point should be at the end of rasterization, to wait for the work of all the threads to end. It is appropriate to place these points at the main thread and to accomplish variable sharing and synchronization through mutex variables. Since creating a new thread is time-consuming, the rendering engine should be implemented in order to avoid continuous thread (re)creation. Threads, for example, can be created after the classification process and ended after their rasterization process, but this will greatly slow down the rendering. Instead, threads live and run continuously. When their work is required, they will be activated, otherwise they will wait. Following these principles the benefit of the implemented renderer is its expected significant performance improvement. The disadvantage at the same time is that the implementation should be more complex.

Although the distributed renderer was emphasized for the rasterization of transparent textures, the principle is also applicable in the case of non-transparent textures. Their rasterization can also be integrated into the model extending the process. As previously mentioned, the rendering process of these types of textures is a set of fast memory copy operations. To keep these properties, the graphical engine should register the type of the textures after the image loading and determine which textures are transparent. During the rasterization process these types of textures are also placed on the same list as the transparent textures. The classification algorithm associates these similarly with one or more logical areas. However, in the actual rendering stage, the engine needs to knowthe exact type of texture, because rasterization of transparent textures is performed by pixel level and that of non-transparent ones is performed most effectively by memory block copy (including the checking and cutting process of the logical boundaries of the areas).

This way the complexity of the rendering engine increases, but it is worth distinguishing the two types in terms of performance.

6 Test Results

The following section presents the performance results of the multi-threaded rasterizer for different test cases. During these tests we considered it important to compare the results to several different solutions. All different test cases have also been implemented by the classical, one-thread rendering solution. In addition, to validate the results the tests were also implemented with the GPU based technology. With this reference value, the relative performance ratio of the methods will be visible and clear.

The GPU based reference implementation was developed with the OpenGL API, where all visual elements were stored in the high-performance video memory and the VBO (Vertex Buffer Object) extension was applied for the rendering. Currently, VBO is the fastest texture rendering method in the GPU area. It is important to emphasize that the drawing of textures was performed using GLSL, where two different test cases were distinguished. The role of the non-optimized type is important in mass texture drawing: a shader object is initialized before every object drawing and closed after it. The marker 'non-optimized' means the cost (performance) of the continuous shader changes. The optimal solution initiates the shader object only once before mass texture drawing and closes it after rendering.

The test programs were written in C++ applying a GCC 4.4.1 compiler and the measurements were performed by an Intel Core i7-870 2.93 GHz CPU. Due to Hyper-Threading technology, the CPU can run eight hardware threads in parallel. As a test environment, a 64 bit Windows 7 Ultimate Edition was chosen. The implementations did not use any hand optimized SSE such as code parts; only the compiler optimized code was applied. The chosen screen resolution and color depth were $800 \times 600 \times 32$ in windowed mode. The hardware used for the test was an ATI Radeon HD 5670 with 1 GB of RAM. To display the software framebuffer, the OpenGL glDrawPixels solution was applied in an optimized form. The alpha-channel images used in the tests contained an average number of transparent pixels, about 50%. During the tests the average *Frame Rate* (Frames Per Second) was recorded for at least one minute run-time. It is important to highlight that in the case of software rendering, the frame rate was 1714 FPS without any drawing, when only the empty framebuffer was sent to the GPU. The pixel operations were optimized for both software renderer solutions (classical and multi-threaded). Framebuffer is defined as an *uint32 t* type array because this storage type makes it possible to handle all the color components of a single pixel in one unsigned integer type variable, in one single block (e.g. color = A << 24 | R << 16 | G << 8 | B) [16].

Table 1 Pixel writing performance

	Speed of rasterization (FPS)	
	Simple rasterizer (1 thread)	Distributed rasterizer (4 thread)
Write 800 × 600 pixels to farmebuffer	614	1122

6.1 Simple Pixel Operation Performance Test

First a very simple but very practical test is worth examining the performance of pixel writing operations. The task of each test implementation was to fill the screen with a predefined color pixel by pixel. Because of the many pixel operations, processing is very computation-intensive. Table 1 summarizes the results.

The results clearly show the advantages of the thread-based solution. The distributed version was almost twice as fast as the classical solution.

6.2 Renderer's Compound Test

In the following our objective is to present and compare the results of the different solutions, applying them to some test cases. Each test represents a special group of tasks. These groups are intended to highlight the most important tasks, those which often occur in computer games. They help to conclude how effective the multi-threaded rasterizer can be in different cases.

Test case 1: during the test we were looking for an answer to the question of how the presented methods can handle a large texture without any transparent areas.

Test case 2: the aim of this test was to measure the renderer's performance applying large and transparent textures. The test image contained an average number of transparent pixels, about 50 %.

Test case 3: the test is a transition between the previous and the following cases. It renders 10 relatively large, non-opaque images.

Test case 4: applying the third test case with transparent textures.

Test case 5: a heavily loaded rendering system was simulated drawing 200 64 × 64 size non-transparent animated objects. Each object has eight different animation frames with identical sizes. Positions are randomly generated and uniformly distributed.

Test case 6: like test case 5 with transparent textures.

Table 2 Benchmark results

Speed of rasterization (FPS)					
	Number of objects	Simple rasterizer (1 thread)	Distributed rasterizer (4 thread)	Non optimized GPU implementation	Optimized GPU implementation
800 × 600 texture (RGB)	1	1580	1710	3012	3012
800 × 600 texture (RGBA)	1	717	1180	3050	3050
256 × 256 texture (RGB)	10	1192	1336	2960	3056
256 × 256 texture (RGBA)	10	522	950	2987	3052
64 × 64 texture (RGB)	200	666	1002	532	1108
64 × 64 texture (RGBA)	200	380	766	538	1126

Table 2 summarizes the results of all solutions.

As we might expect, pixel level rasterization has the lowest performance in all cases. While the rendering performance of the non-opaque textures is higher, in the transparent case it was much worse. The reason for this is that the renderer draws non-opaque textures with memory copy operations, and transparent textures pixel by pixel. The performance values achieved underline the fact that a rasterizer using one thread is not able to exploit the available CPU resources.

The presented distributed renderer engine performed well in all cases. Although the implemented four- thread based prototype still does not properly take into account the hardware cores, the results are convincing. In one case, its performance was higher than for the non-optimized GPU solution. In addition, it should be noted that this approach scores better on the high graphical load. While in the second test case the rate of performance values of optimized GPU and the distributed approach was 2.58, the rate is 2.09 in the last test case.

The optimized GPU implementation has the fastest performance in all cases. But we should not forget that the calculations are performed by dedicated hardware. There is no need to move data between the GPU and the main memory.

Naturally, in practice there could be additional (exceptional) cases: one example would be in a game if all the objects are positioned in one logical area. In this case one render thread will render all the objects and its performance will be the same as that of the classical approach. Besides, the above examples do not take into account the case where the image of an object should be scaled or rotated. Compared to the GPU based implementation, this requires more resources from the CPU.

7 Conclusion and Further Work

Although today the field of computer graphics is dominated by the GPU market, we cannot forget the opportunities offered by software based image synthesis. The central units have undergone a huge revolution during the recent years, offering new opportunities in this area. A powerful GPU cannot be defeated in rasterization performance, but a properly designed software renderer based on modern concepts and solutions is also able to achieve good results in rasterization, not only in speed, but also in flexibility. It should not be forgotten that a fully software based pipeline is less restricted compared to today's hardware solutions. In addition, the model discussed in this chapter highlights that there are grounds also for developing two-dimensional games and other graphics applications using software renderers.

Further development of central processing units (e.g. the AVX instruction set) will open up more and more opportunities in this area. Naturally this will require a great deal of effort and applying lower level languages (e.g. C, C++, D) that can take advantage of these potentials of the central unit.

In further work we would like to find the answer to the question of to what extent the performance of the rasterization process can be enhanced by utilizing the CPU's SIMD extensions in both 2D and 3D cases.

Acknowledgments The described work was carried out as part of the TÁMOP-4.2.2/B-10/1-2010-0008 project in the framework of the New Hungarian Development Plan. The realization of this project is supported by the European Union, co-financed by the European Social Fund.

References

1. Zach, B.: A Modern approach to software rasterization. University Workshop, Taylor University (2011)
2. TransGaming Inc: Swiftshader: Why the future of 3D Graphics is in Software. Technology white paper. Developer site: http://transgaming.com. Accessed 29 Jan 2013
3. Microsoft Corporation: Windows advanced rasterization platform (WARP) guide. MSDN library: DirectX Graphics Articles. Developer site: http://msdn.microsoft.com. Accessed 22 Jan 2013
4. Abrash, M.: Rasterization on larrabee, Intel developer site: http://software.intel.com/en-us/articles/rasterization-on-larrabee. Accessed 20 Feb 2013
5. Seiler, L., Carmean, D., Sprangle, E., Forsyth, T., Abrash, M., Dubey, P., Junkins, S., Lake, A., Sugerman, J., Cavin, R., Espasa, R., Grochowski, E., Juan, T., Hanrahan, P.: Larrabee: a many-core x86 architecture for visual computing. In: ACM Transactions on Graphics (TOG)—Proceedings of ACM SIGGRAPH 27 Aug 2008
6. Rost, R.: The OpenGL shading language. Addison Wesley, London (2004)

7. Laine, S., Karras, T.: High-performance software rasterization on GPUs. High Performance Graphics, Vancouver (2011)
8. Akenine-Möller, T., Haines, E.: Real-time rendering. 3rd edn, A. K. Peters/CRC Press, Wellesley (2008)
9. Sugerman, J., Fatahalian, K., Boulos, S., Akeley, K., Hanrahan, P.: Gramps: a programming model for graphics pipelines. ACM Trans. Graph. **28**, 1–11 (2009)
10. Fang, L., Mengcheng, H., Xuehui, L., Enhua, W.: FreePipe: a programmable, parallel rendering architecture for efficient multi-fragment effects. In: Proceedings of ACM SIGGRAPH Symposium on Interactive 3D Graphics and Games (2010)
11. Agner, F.: Optimizing software in C++: An optimization guide for Windows, Linux and Mac platforms. Study at Copenhagen University College of Engineering. Paper site: http://www.agner.org/optimize. (2012)
12. Swenney, T.: The end of the GPU roadmap. In: Proceedings of the Conference on High Performance Graphics, pp. 45–52 (2009)
13. Coffin, C.: SPU-based deferred shading for battlefield 3 on playstation 3. In: Game Developer Conference Presentation 8 (2011)
14. RAD Game Tools: Pixomatic advanced software rasterizer. http://www.radgametools.com/pixomain.htm (2012)
15. Akhter, S., Roberts, J.: Multi-core programming—increasing performance through software multi-threading, 1st edn. Intel Corporation (2006)
16. Mileff, P., Dudra, J.: Efficient 2D software rendering. Prod Syst Inform Eng **6**, 99–110 (2012)
17. Loop, C., Eisenacher, C.: Real-time patch-based sort-middle rendering on massively parallel hardware. Microsoft Research, Tech. Rep., MSR-TR-2009-83 (2009)

Numerical Analysis of Free Convection from a Vertical Surface Embedded in a Porous Medium

Krisztián Hriczó and Gabriella Bognár

Abstract The numerical solutions for free convective heat transfer in a viscous fluid flow over a vertical flat plate embedded in a porous medium under mixed thermal boundary conditions are examined. The governing equations are derived assuming the linear Darcy model and the nonlinear density temperature variation in the buoyancy force term. Applying a similarity transformation the transformed system of ordinary differential equations is investigated numerically.

List of Symbols

g	Acceleration due to gravity
K	Permeability of the porous medium
L	Reference length
n, m	Similarity exponent
Nu	Nusselt number
Ra	Rayleigh number
T	Fluid temperature
T_r	Reference temperature
T_∞	Ambient temperature
$\bar{u}, \bar{v}$	Dimensional velocity components along $\bar{x}$ and $\bar{y}$
u, v	Non-dimensional velocity components
$\bar{x}, \bar{y}$	Dimensional Cartesian coordinates
x, y	Non-dimensional variables

K. Hriczó (✉) · G. Bognár
Department of Analysis, University of Miskolc, Miskolc-Egyetemváros 3515, Hungary
e-mail: krisztian.hriczo@gmail.com

G. Bognár and T. Tóth (eds.), *Applied Information Science, Engineering and Technology*, Topics in Intelligent Engineering and Informatics 7, DOI: 10.1007/978-3-319-01919-2_6,

Greek Symbols

α_m	Effective thermal diffusivity
β_1, β_2	Coefficients of thermal expansion
δ	Parameter
ϵ	Mixed thermal boundary condition parameter
η	Similarity variable
ν	Kinematic viscosity
ρ	Density of fluid
ρ_∞	Density of ambient fluid
Θ	Non-dimensional temperature
Ψ	Non-dimensional stream function

1 Introduction

The study of convective heat transfer from surfaces embedded in a porous medium has attracted considerable interest in the past several decades. The interest in this field is due to the numerous applications of flow through a porous medium, for instance moisture transport in grain storage installations, filtration, transpiration cooling, oil extraction, ground water pollution, thermal insulation and geothermal systems.

A porous medium can be considered to be a material consisting of a solid matrix with interconnected voids (pores) and the solid matrix can be either rigid (the usual configuration) or it may undergo some small deformation. The interconnectedness of the voids allows the flow of a single fluid, or of multiple fluids through the material. Some natural porous media are sandstone, wood and limestone [7]. A typical oil reservoir is a body of underground rock in which there exists an interconnected void space occupying up to 30 % of the bulk volume. This void space harbors oil, brine and injected fluids. The more sophisticated oil recovery technologies require developing mathematical reservoir models, testing various operating strategies and comparing different recovery technologies [1]. There is a need to characterize the flow around a deep geological repository for the disposal of high-level nuclear waste, e.g. spent fuel rods from nuclear reactors [7, 8].

Free or mixed boundary layer convection flow over horizontal or vertical surfaces with constant or variable wall temperatures has been widely investigated. Reviews on convection flow through porous media have been published by Nield and Bejan [10] and in [6]. Theoretical and experimental work on this subject have been presented in [3–15]. Free convection from a vertical impermeable surface in a saturated porous medium has many important geophysical and engineering applications. For example, hot dike complexes in a volcanic region can provide an energy source for the heating of ground-water (see Furumoto [5]) which can be used for power generation. From a technological aspect, the analysis of the free convection of fluids with known physical properties is important.

Cheng and Minkowycz [3] analyzed the problem of natural convection heat transfer with constant viscosity from a vertical flat plate embedded in a porous medium, where the prescribed wall temperature is a power function of height. They applied Boussinesq's approximation for the density temperature variation and assumed that the convection took place in a thin layer around the heating surface. They observed that the convection took place within a thin boundary layer around the heated surface. Cheng and Chang [4] studied the flow properties from a horizontal surface, assuming also that the temperature of the flat plate varied with a power of the distance from the leading edge. The case of a flat plate with constant heat flux was considered by Rees and Pop [13].

Nazar et al. [9] studied the free convection boundary layer over vertical and horizontal surfaces embedded in a fluid saturated porous medium with mixed thermal boundary conditions assuming the linear density temperature variation (Boussinesq's approximation).

In the above-mentioned papers the physical properties of the fluid were assumed to be constant. However, it is known that some fluid properties, such as density, viscosity and thermal conductivity, vary with temperature. In order to give a better description of the flow and temperature characteristics, the influence of the variable fluid properties should be considered. When the temperature difference between the surface and the ambient fluid is relatively large, the nonlinear density temperature variation in the buoyancy force term may exert a strong influence on the fluid flow characteristics (see [2, 14–16]). This typical phenomenon appears e.g., in groundwater in vertical fissures or cracks during the natural convection currents or in residual warm water discharged from a geothermal power plant.

This chapter examines the solutions for free convective heat transfer in a viscous fluid flow over vertical flat plates embedded in a porous medium under mixed thermal boundary conditions. The governing equations are derived using the Darcy model and assuming nonlinear density temperature variation in the buoyancy force term. Applying a similarity transformation, the system of partial differential equations is transformed to ordinary differential equations and the resulting coupled, nonlinear ordinary differential equations are investigated analytically and numerically. The aim is to extend the work of Nazar et al. [9] by investigating the convective boundary layers on an impermeable vertical plate.

2 Basic Equations for Boundary Layers

Let us consider a vertical flat plate in a porous medium. The steady free convection on a flat plate embedded in a fluid-saturated porous medium of uniform temperature T_∞ will be investigated. The thermal and momentum boundary layers are exemined along a flat surface. Dimensional coordinates are used with the $\bar{x}$-axis measured along the surface and the $\bar{y}$-axis being normal to it. The velocity components along the $\bar{x}$- and $\bar{y}$-axes are denoted by $\bar{u}(\bar{x}, \bar{y})$ and $\bar{v}(\bar{x}, \bar{y})$, respectively (see Fig. 1).

First, the governing equations on the boundary-layer are introduced. The following assumptions are made [11, 15]:

(i) The convective fluid and the porous medium are in local thermodynamic equilibrium;

Fig. 1 Physical model

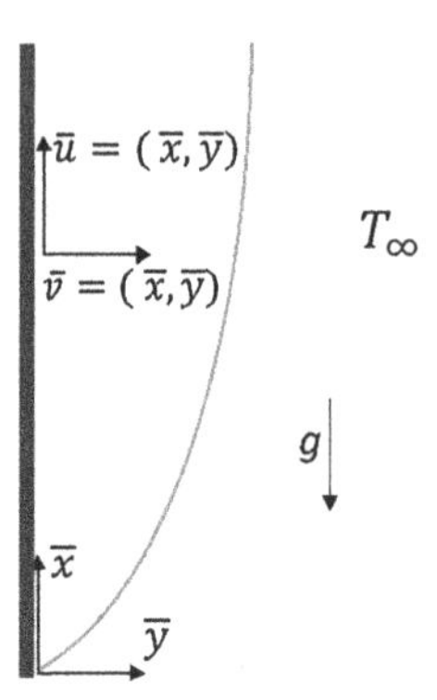

(ii) the temperature of the fluid is below the boiling point;
(iii) the convective flow is due to the density difference between the source (or sink) and that at infinity;
(iv) the viscosity, thermal conductivity and permeability of the fluid and the porous medium are constant.

If the flow is steady and the fluid density is constant the continuity of mass equation is given by

$$\frac{\partial \bar{u}}{\partial \bar{x}} + \frac{\partial \bar{v}}{\partial \bar{y}} = 0. \tag{1}$$

In case of flow through porous medium, the Navier-Stokes equation for conservation of momentum is replaced by Darcy's law. We shall assume that the density is constant except in the body force term. For small changes in temperature, when thermal effects are considered and the buoyancy forces are comparable with inertial and viscous forces, the Boussinesq approximation is applied:

$$\rho = \rho_\infty[1 - \beta_1(T - T_\infty)], \tag{2}$$

where ρ is the density of the fluid, T is the temperature of the fluid, ρ_∞ is the density at the ambient temperature T_∞ and β_1 refers to the thermal expansion coefficient of the fluid. When the temperature difference between the surface and the ambient fluid is large enough, the nonlinear density temperature variation in the buoyancy force term is applied (see [3, 15, 16])

$$\rho = \rho_\infty[1 - \beta_2(T - T_\infty)^2], \tag{3}$$

where β_2 is the coefficient of the thermal expansion of the second order. The relations in Eqs. (2) and (3) accommodate the linear density temperature variation and the quadratic density temperature variation, respectively. For both relations, the equation governing the fluid flow can be written as

$$\bar{u} = \frac{gK\beta_\delta}{\nu}(T - T_\infty)^\delta, \tag{4}$$

where g denotes the acceleration due to gravity, K is the permeability of the porous medium, ν denotes kinematic viscosity and $\delta = 1$ stands for the linear density temperature variation in Eq. (2) while $\delta = 2$ for the nonlinear case in Eq. (3).

Energy conservation can be expressed by

$$\bar{u}\frac{\partial T}{\partial \bar{x}} + \bar{v}\frac{\partial T}{\partial \bar{y}} = \alpha_m \frac{\partial^2 T}{\partial y^2}, \tag{5}$$

where α_m is the effective thermal diffusivity of the fluid-saturated porous medium.

In general, the non-dimensional variables are defined as follows

$$x = \bar{x}/L, \tag{6}$$

$$y = Ra^{1/2}(\bar{y}/L), \tag{7}$$

$$u = Ra^{-1}(L/\alpha_m)\bar{u}, \tag{8}$$

$$v = Ra^{-1/2}(L/\alpha_m)\bar{v}, \tag{9}$$

$$\theta = (T - T_\infty)/(T_r - T_\infty), \tag{10}$$

where L is a characteristic length of the plate, T_r is the reference temperature and for the porous medium

$$Ra = gK\beta_1(T_r - T_\infty)L/\alpha_m\nu$$

is the Rayleigh number for the case $\delta = 1$ and

$$Ra = gK\beta_2(T_r - T_\infty)^2 L/\alpha_m\nu$$

for $\delta = 2$. In terms of the non-dimensional variables defined by Eqs. (6)–(10), the governing equations (1), (4) and (5) of the boundary layer can be written as the system of the following equations (see Pop and Ingham [12])

$$\frac{\partial u}{\partial x} + \frac{\partial v}{\partial y} = 0, \tag{11}$$

$$u = \theta^\delta, \tag{12}$$

$$u\frac{\partial \theta}{\partial x} + v\frac{\partial \theta}{\partial y} = \frac{\partial^2 \theta}{\partial y^2}, \tag{13}$$

which are to be solved with the following boundary conditions:
at the plate ($y = 0$)

$$v\,(x,0)=0, \tag{14}$$

$$A(x)(T_r - T_\infty)\theta(x,0) - B(x)(T_r - T_\infty)\frac{Ra^{1/2}}{L}\left(\frac{\partial\theta}{\partial y}\right)_{y=0} = C(x), \tag{15}$$

and far from the plate ($y \to \infty$)

$$\theta(x,y) \to 0, \tag{16}$$

where $A(x)$, $B(x)$ and $C(x)$ are undetermined functions of x.

To study this problem, it is convenient to introduce the stream function ψ defined by

$$u = \frac{\partial\psi}{\partial y},\ v = -\frac{\partial\psi}{\partial x}. \tag{17}$$

Then, the continuity equation (11) is satisfied automatically and Eqs. (12) and (13) become

$$\frac{\partial\psi}{\partial y} = \theta^\delta, \tag{18}$$

$$\frac{\partial\psi}{\partial y}\frac{\partial\theta}{\partial x} - \frac{\partial\psi}{\partial x}\frac{\partial\theta}{\partial y} = \frac{\partial^2\theta}{\partial y^2}, \tag{19}$$

The initial condition (14) can be written as

$$\frac{\partial\psi}{\partial x}\,(x,0) = 0.$$

Now, we look for similarity solutions to Eqs. (18) and (19) of the following form

$$\psi(x,y) = x^p f\,(\eta)\,, \quad \theta(x,y) = x^q h\,(\eta)\,, \quad \eta = x^r y. \tag{20}$$

Substituting (20) into Eqs. (18) and (19), one gets

$$x^{p+r} f' = (x^q h)^\delta, \tag{21}$$

$$x^{p-1}(qf'h - pfh') = x^r h'', \tag{22}$$

where primes denote differentiation with respect to η.

We distinguish two cases.

Case 1: ($\delta = 1$) By setting $p = 1 + n$, $q = 1 + 2n$ and $r = n$, Eqs. (21) and (22) become

$$f' = h, \tag{23}$$

$$h'' - (1+2n)\, f'h + (1+n) f h' = 0. \tag{24}$$

The boundary conditions (14)–(16) can be written as
$\eta = 0:$

$$f\,(0) = 0, \tag{25}$$

$$a(x)(T_r - T_\infty)h(0) - b(x)(T_r - T_\infty)^{3/2}h'(0) = 1, \tag{26}$$

as $\eta \to \infty:$

$$h(\eta) \to 0, \tag{27}$$

where functions $a(x)$ and $b(x)$ are given by

$$a(x) = \frac{A(x)}{C(x)} x^{1+2n}, \quad b(x) = \frac{B(x)}{C(x)} x^{1+3n} \left(\frac{gK\beta}{\alpha_m \nu L}\right)^{1/2}. \tag{28}$$

In order to have similarity solutions to Eqs. (23) and (24) subject to boundary conditions (25)–(27), these functions must be equal to a constant. Without loss of generality, we choose the constants a, b and T_∞ such that for the reference temperature T_r we get:

$$b(T_r - T_\infty)^{3/2} + a(T_r - T_\infty) = 1. \tag{29}$$

When $\epsilon = b(T_r - T_\infty)^{3/2}$, the thermal boundary condition (26) has the form

$$(1-\epsilon)h(0) - \epsilon h'(0) = 1. \tag{30}$$

We note that the case $\epsilon = 0$ corresponds to prescribed surface temperature with $\theta(x, 0) = x^{1+2n}$, the case $\epsilon = 1$ corresponds to prescribed surface heat flux with $(\partial\theta/\partial y)_{y=0} = -x^{1+3n}$ and the case $\epsilon \to \infty$ corresponds to the mixed boundary condition $(\partial\theta/\partial y)_{y=0} = -x^n\theta(x, 0)$.
Case 2: ($\delta = 2$) By setting $p = 1+m$, $q = (1+2m)/2$ and $r = m$, Eqs. (21) and (22) can be formulated as follows

$$f' = h^2, \tag{31}$$

$$h'' - \frac{1+2m}{2} f'h + (1+m) f h' = 0. \tag{32}$$

The boundary conditions (14)–(16) are reduced to the following forms at $\eta = 0$:

$$f(0) = 0, \tag{33}$$

$$a(x)(T_r - T_\infty)h(0) - b(x)(T_r - T_\infty)^2 h'(0) = 1, \tag{34}$$

as $\eta \to \infty$:

$$h(\eta) \to 0, \tag{35}$$

where functions $a(x)$ and $b(x)$ are given by

$$a(x) = \frac{A(x)}{C(x)} x^{\frac{1+2m}{2}}, \quad b(x) = \frac{B(x)}{C(x)} x^{\frac{1+4m}{2}} \left(\frac{gK\beta}{\alpha_m \nu L} \right)^{1/2}. \tag{36}$$

In order that similarity solutions to Eqs. (31) and (32) with boundary conditions (33)–(35) exist, functions $a(x)$ and $b(x)$ must be equal to a constant. For given values of constants a, b and T_∞, the reference temperature T_r can be chosen to satisfy the following equation:

$$b(T_r - T_\infty)^2 + a(T_r - T_\infty) = 1. \tag{37}$$

Let us define $\epsilon = b(T_r - T_\infty)^2$. Then the thermal boundary condition (34) can be formulated as follows

$$(1 - \epsilon)h(0) - \epsilon h'(0) = 1. \tag{38}$$

It should be noted that the case $\epsilon = 0$ corresponds to prescribed surface temperature $\theta(x, 0) = x^{(1+2m)/2}$, the case $\epsilon = 1$ corresponds to prescribed surface heat flux $(\partial\theta/\partial y)_{y=0} = -x^{1/2+2m}$ and the case $\epsilon \to \infty$ corresponds to the mixed boundary condition $(\partial\theta/\partial y)_{y=0} = -x^m \theta(x, 0)$.

3 Exact Solution

When the Boussinesq approximation is applied, we note that the special case $n = 0$ admits an explicit solution. Solutions to the system of ordinary differential equations (23), (24) for with boundary conditions (25), (27) and $h(0) = -h'(0)$ ($\epsilon \to \infty$) in closed form can be given as

$$f(\eta) = 1 - e^{-\eta}, \quad h(\eta) = e^{-\eta}.$$

Due to the inherent complexity of such flows, in general it is almost impossible to provide exact analytical solutions.

4 Results

The nonlinear ordinary differential equations (23), (24) subject to the boundary conditions in (25)–(27) were solved numerically by Nazar et al. [9] for different values of the parameters n using the Keller-box method. We have solved the two boundary value problems, Eqs. (23), (24) subject to the boundary conditions (25)–(27) and Eqs. (31), (32) subject to the boundary conditions (33)–(35), by using the symbolic algebra software Maple 12. For linear density temperature variation ($\delta = 1$) some values of $-h'(0)$ for several values of the parameter n in the range $-2/3 \leq n \leq 1$ are given in Table 1 for $\epsilon = 0$, the values of $h(0)$ are shown in Table 2 for $\epsilon = 1$ and $h(0) = -h'(0)$ in Table 3 for $\epsilon \to \infty$; they are compared with those reported by Nazar et al. [9]. It can be seen that the results are in excellent agreement.

For quadratic density temperature variation ($\delta = 2$), the numerical results for the three cases ($\epsilon = 0$, $\epsilon = 1$, $\epsilon \to \infty$) are exhibited in Table 4 for parameter values $-2/3 \leq m \leq 100$. In Tables 5, 6 and 7, the values of $-h'(0)$ for $\epsilon = 0$, $h(0)$ for $\epsilon = 1$ and $h(0) = -h'(0)$ for $\epsilon \to \infty$ are compared when $\delta = 1$ or $\delta = 2$.

Figures 2, 3 and 4 exhibit the Maple generated numerical values of $-h'(0)$ ($\epsilon = 0$), $h(0)$ ($\epsilon = 1$) and $h(0) = -h'(0)$ ($\epsilon \to \infty$) for different values of n and m for linear and quadratic density temperature variations. From Fig. 3 we conjecture that $h(0) \to \infty$ as $n,\ m \to -1 + 0$, and from Fig. 4 we find that for some values of n and m, $h(0) = h'(0) \to \infty$ as well.

In all cases, our numerical results show a decreasing trend, which is a good agreement with those reported by Nazar et al. [9]. In Figs. 5, 6, 7, 8, 9 and 10 it is noticed that as n or m increases, the velocity and temperature profiles decrease.

Table 1 Values of $-h'(0)$ for prescribed surface temperature ($\epsilon = 0$) and linear density temperature variation ($\delta = 1$)

$f' = h$	Nazar et al. [9]	Our results
n	$-h'(0)(\epsilon = 0)$	
−2/3	0.0000	0.00000
−5/8	0.1620	0.16203
−1/2	0.4437	0.44374
−3/8	0.6266	0.62655
−1/3	0.6776	0.67764
−1/4	0.7704	0.77036
−1/8	0.8923	0.89234
0	1.0000	1.00000
0.1	1.0786	1.07858
0.2	1.1519	1.15191
0.3	1.2209	1.22090
0.4	1.2862	1.28624
0.5	1.3484	1.34845
1	1.6443	1.64213

Table 2 Values of $h(0)$ for prescribed surface heat flux ($\epsilon = 1$) and linear density temperature variation ($\delta = 1$)

$f' = h$	Nazar et al. [9]	Our results
n	$h(0)(\epsilon = 1)$	
−2/3	–	–
−5/8	3.3645	3.36454
−1/2	1.7188	1.71886
−3/8	1.3657	1.36571
−1/3	1.2962	1.29617
−1/4	1.1900	1.18996
−1/8	1.0789	1.07889
0	1.0000	0.99999
0.1	0.9508	0.95081
0.2	0.9100	0.91002
0.3	0.8754	0.87540
0.4	0.8455	0.84550
0.5	0.8193	0.81929
1	0.7237	0.72367

Table 3 Values of $h(0)$ for mixed boundary condition ($\epsilon \to \infty$) and linear density temperature variation ($\delta = 1$)

$f' = h$	Nazar et al. [9]	Our results
n	$h(0) = -h'(0)(\epsilon \to \infty)$	
−2/3	–	–
−5/8	–	–
−1/2	–	–
−3/8	1.9080	2.53540
−1/3	1.8007	2.16579
−1/4	1.6735	1.67379
−1/8	1.2457	1.24596
0	0.9912	0.99136
0.1	0.8516	0.85179
0.2	0.7445	0.74564
0.3	0.6626	0.66438
0.4	0.5968	0.59847
0.5	0.5411	0.54444
1	0.3710	0.37500

The physical quantity of interest in this problem is the Nusselt number, Nu, which can be expressed as

$$Nu = -x^{1+3n}\sqrt{Ra}\, h'(0) \quad \text{as} \quad \delta = 1$$

Table 4 Values of $-h'(0)$ ($\epsilon = 0$), $h(0)$ ($\epsilon = 1$) and $h(0) = -h'(0)$ ($\epsilon \to \infty$) for quadratic density temperature variation ($\delta = 2$)

$f' = h^2$			
m	$-h'(0)(\epsilon = 0)$	$h(0)(\epsilon = 1)$	$h(0) = -h'(0)(\epsilon \to \infty)$
−3/4	0.00004	–	–
−2/3	0.18029	2.35499	5.51375
−13/20	0.20572	2.20387	4.86134
−31/48	0.21177	2.17212	4.72248
−5/8	0.24042	1.89620	4.14615
−1/2	0.37653	1.62920	2.64990
−3/8	0.47946	1.44378	2.07758
−1/3	0.50952	1.40055	1.95417
−1/4	0.56514	1.32988	1.76064
−1/8	0.63993	1.24979	1.55368
0	0.70710	1.18898	1.40533
0.1	0.75665	1.14941	1.31287
0.2	0.80321	1.11562	1.23644
0.3	0.84726	1.08625	1.17191
0.4	0.88915	1.06037	1.11650
0.5	0.92918	1.03729	1.06824
1	1.10798	0.94995	0.89539
1.1	1.14041	0.93635	0.87687
1.3	1.20264	0.91181	0.83149
1.5	1.26182	0.89018	0.79250
2	1.39890	0.84545	0.71484
3	1.63907	0.78107	0.61010
4	1.84833	0.73553	0.54102
5	2.03621	0.70078	0.49110
10	2.79200	0.59846	0.35815
20	3.88522	0.50733	0.25738
50	6.08222	0.40547	0.16441
100	8.57268	0.34154	0.11664

and

$$Nu = -x^{1/2+2m}\sqrt{Ra}\,h'(0) \quad \text{as } \delta = 2.$$

As a consequence of our calculations, we find that for both linear and quadratic density temperature variations, increasing the parameters n and m leads to an increase in the Nusselt number. Figures 5, 7 and 9 describe the behavior of $f'(\eta)$ for $\delta = 1$ and $\delta = 2$. We notice that the boundary layer thickness increases as n and m decrease. Moreover, the boundary layer thickness is greater for $\delta = 2$ than for $\delta = 1$.

Figure 6 demonstrates that the thermal boundary layer thickness is greater for $\delta = 2$ than for $\delta = 1$ taking the same values of n and m. As $u = x^{1+2n} f'$ and $u = x^{1+2m} f'$ Fig. 7 displays the velocity profiles for linear and quadratic density temperature variation for the prescribed heat flux. The velocity component f' at

Table 5 Comparison of the values $-h'(0)$ for $\delta = 1$ and $\delta = 2$ when $\epsilon = 0$

	$f' = h$	$f' = h^2$
n/m	$-h'(0)(\epsilon = 0)$	
−2/3	0.00000	0.18029
−5/8	0.16203	0.24042
−1/2	0.44374	0.37653
−3/8	0.62655	0.47946
−1/3	0.67764	0.50952
−1/4	0.77036	0.56514
−1/8	0.89234	0.63993
0	1.00000	0.70710
0.1	1.07858	0.75665
0.2	1.15191	0.80321
0.3	1.22090	0.84726
0.4	1.28624	0.88915
0.5	1.34845	0.92918
1	1.64213	1.10798

Table 6 Comparison of the values $h(0)$ for $\delta = 1$ and $\delta = 2$ when $\epsilon = 1$

	$f' = h$	$f' = h^2$
n/m	$h(0)(\epsilon = 1)$	
−2/3	–	2.35499
−5/8	3.36454	1.89620
−1/2	1.71886	1.62920
−3/8	1.36571	1.44378
−1/3	1.29617	1.40055
−1/4	1.18996	1.32988
−1/8	1.07889	1.24979
0	0.99999	1.18898
0.1	0.95081	1.14941
0.2	0.91002	1.11562
0.3	0.87540	1.08625
0.4	0.84550	1.06037
0.5	0.81929	1.03729
1	0.72367	0.94995

zero decreases as n or m increase both in linear or nonlinear cases. Increasing the parameter n or m leads to a decrease in the boundary layer thickness. It is also found from Fig. 8 that the wall temperature decreases while n or m increases. For mixed thermal boundary condition the hydrodynamic and thermal boundary layer thicknesses are thinner than in the cases of $\epsilon = 0$ and $\epsilon = 1$. Increasing n or m leads to a decrease in $f'(0)$ (see Fig. 9). Figure 10 displays the temperature profiles for $\epsilon \to \infty$, and it can be observed that the convection takes place in a thin boundary

Table 7 Comparison of the values $h(0)$ for ($\delta = 1$ and $\delta = 2$ when $\epsilon \to \infty$

	$f' = h$		$f' = h^2$
n/m		$h(0) = -h'(0)(\epsilon \to \infty)$	
−2/3	–		5.51375
−5/8	–		4.14615
−1/2	–		2.64990
−3/8	2.53540		2.07758
−1/3	2.16579		1.95417
−1/4	1.67379		1.76064
−1/8	1.24596		1.55368
0	0.99136		1.40533
0.1	0.85179		1.31287
0.2	0.74564		1.23644
0.3	0.66438		1.17191
0.4	0.59847		1.11650
0.5	0.54444		1.06824
1	0.37500		0.89539

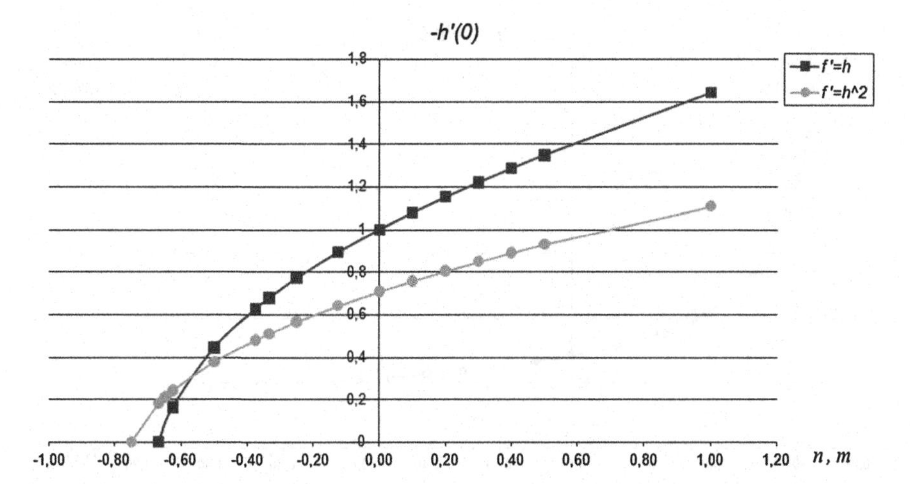

Fig. 2 Variation of $-h'(0)$ for linear and quadratic density temperature variations when $\epsilon = 0$

layer around the heating surface. Figures 5, 6, 7, 8, 9 and 10 demonstrate the velocity and temperature profiles and show similar patterns for $\delta = 1$ and $\delta = 2$. From the nonlinear density temperature variation exhibited in Figs. 7 and 8 it can be seen that for $\delta = 2$, $f'(0)$ and $h(0)$ are smaller than for $\delta = 1$ as $\epsilon = 1$ taking the same values of n and m. The effect of δ is opposite on $f'(0)$ and $h(0)$ for $\epsilon \to \infty$.

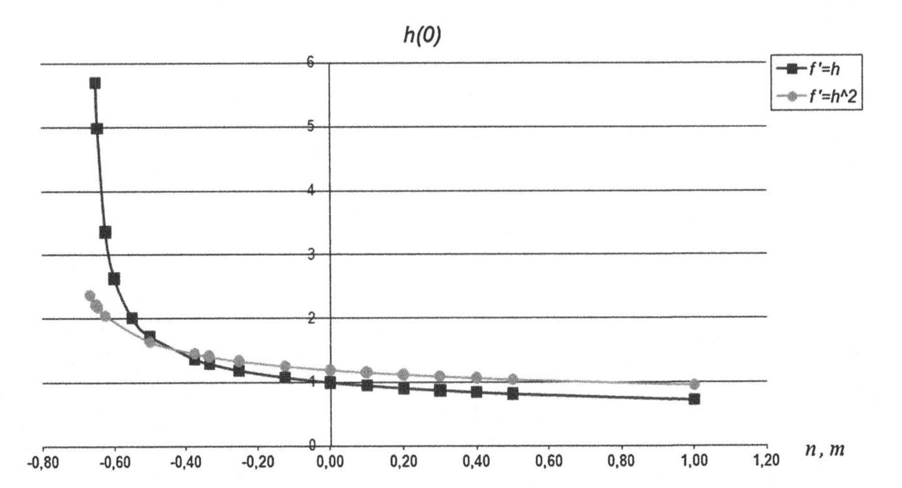

Fig. 3 Variation of $h(0)$ for linear and quadratic density temperature variations when $\epsilon = 1$

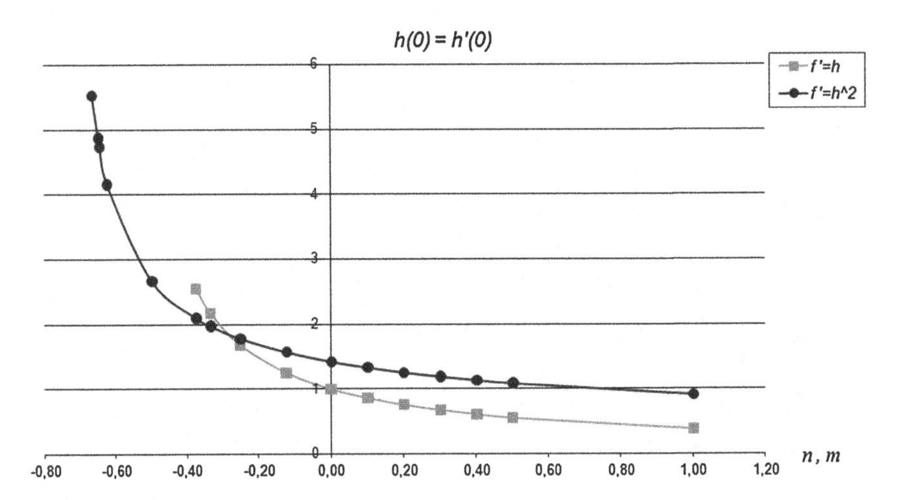

Fig. 4 Variation of $h(0)$ for linear and quadratic density temperature variations when $\epsilon \to \infty$

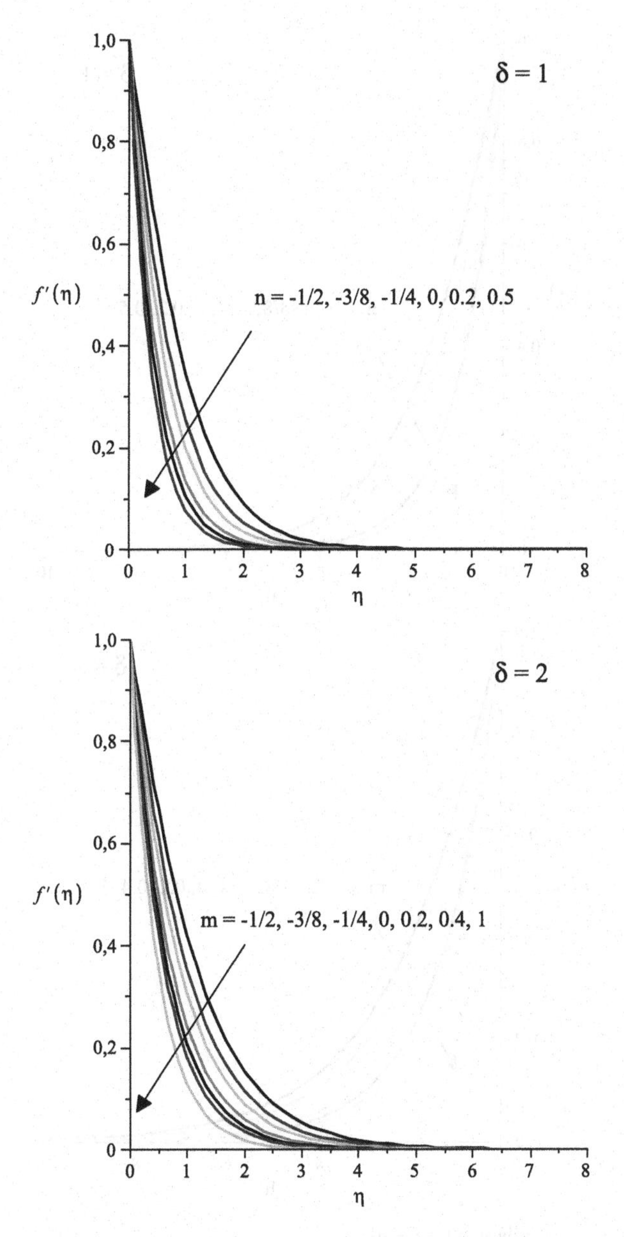

Fig. 5 Velocity distribution for $\epsilon = 0$

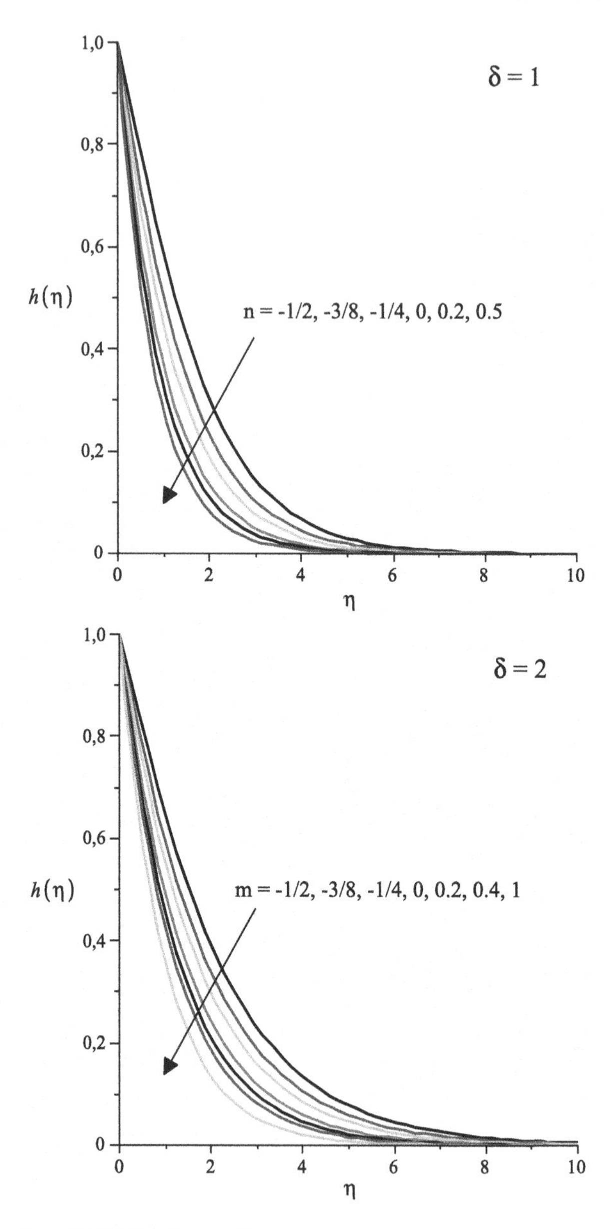

Fig. 6 Temperature distribution for $\epsilon = 0$

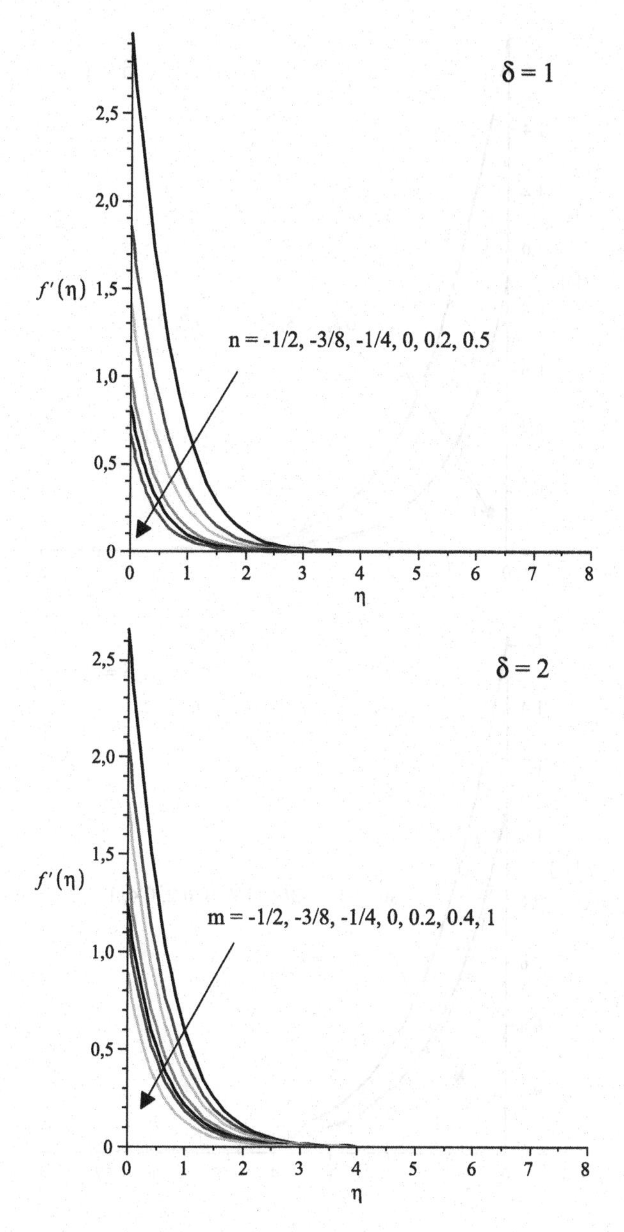

Fig. 7 Velocity distribution for $\epsilon = 1$

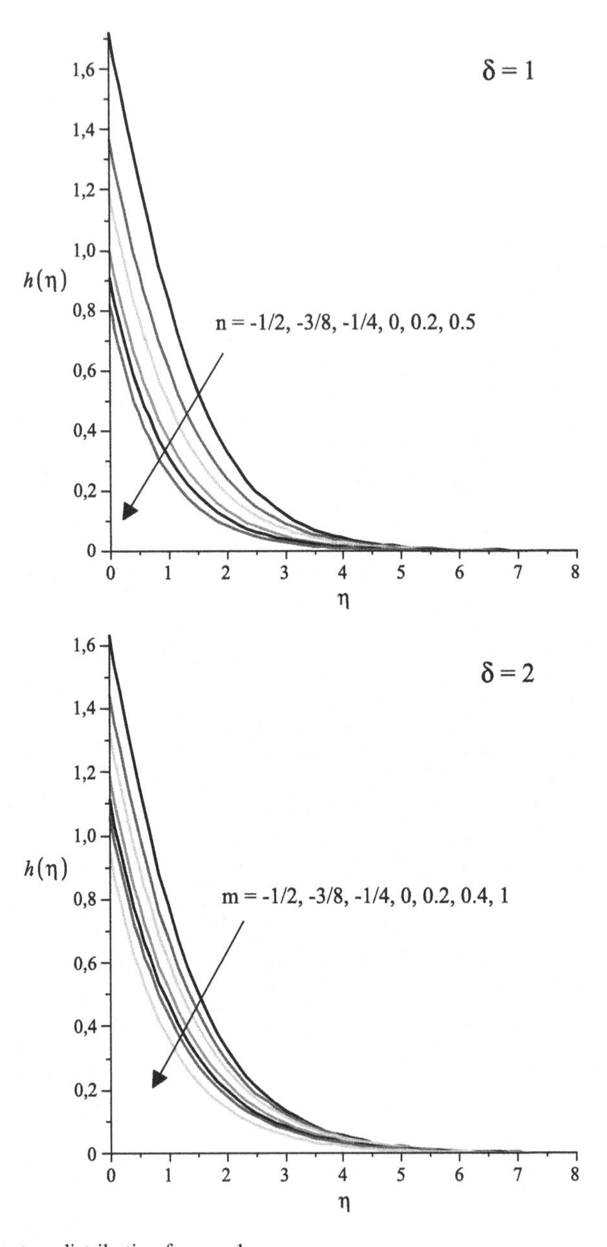

Fig. 8 Temperature distribution for $\epsilon = 1$

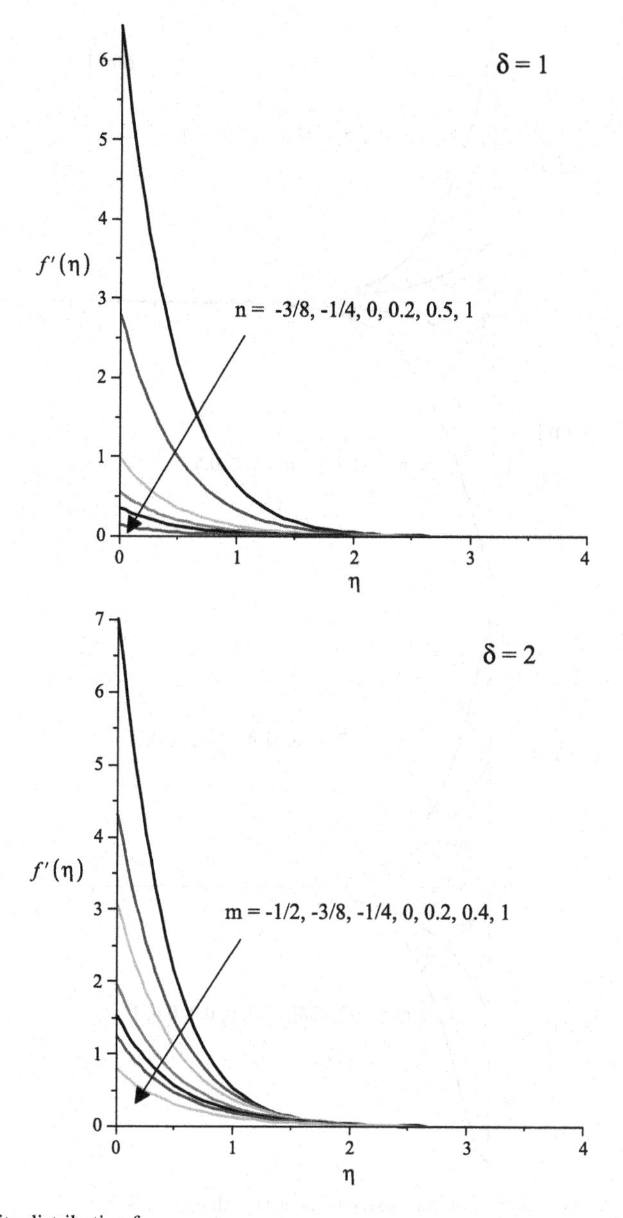

Fig. 9 Velocity distribution for $\epsilon \to \infty$

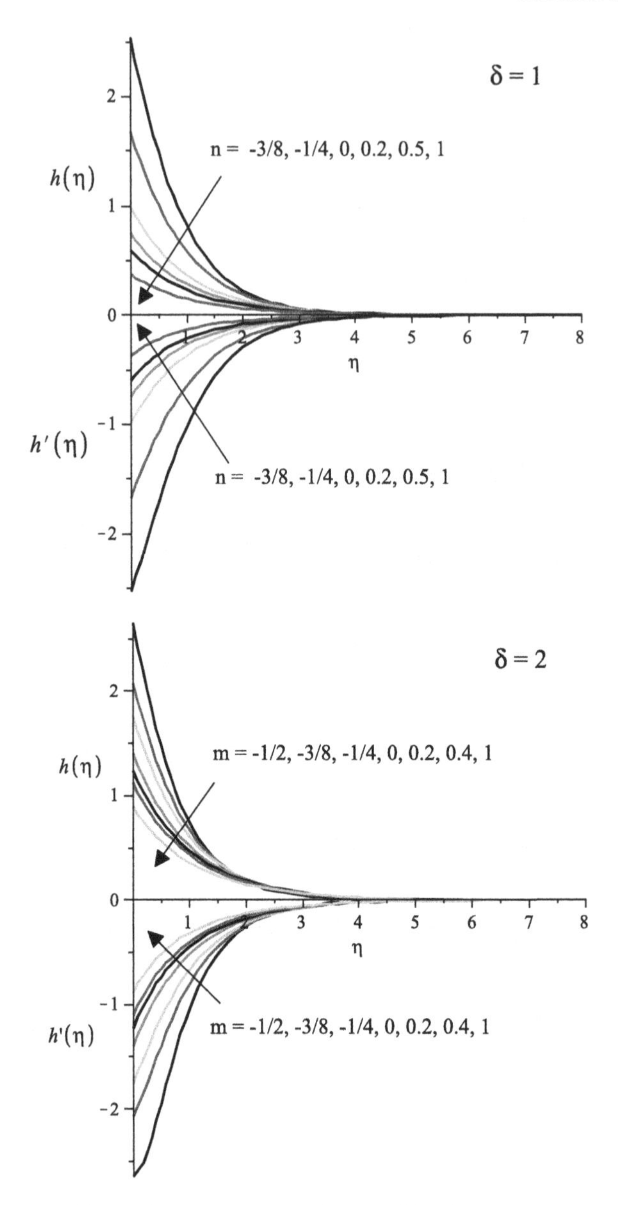

Fig. 10 Temperature distribution for mixed boundary condition ($\epsilon \to \infty$)

5 Conclusions

The free convection flow from a vertical surface embedded in a porous media is examined when the temperature difference between the surface and the fluid is supposed to be small or relatively large. In these two cases, linear or quadratic density temperature variations are applied. Using the similarity transformation method the basic equations are reduced to nonlinear ordinary differential equations. The numerical solutions are compared for linear density temperature variation with those reported by Nazar et al. [9]. We have found excellent agreement between the obtained data. For nonlinear density temperature variation the numerical results are also exhibited in the chapter. We presented figures in order to exhibit the temperature and velocity profiles for both linear and nonlinear cases.

Acknowledgments The described work was carried out as part of the TÁMOP-4.2.2/B-10/1-2010-0008 project in the framework of the New Hungarian Development Plan. The realization of this project is supported by the European Union, co-financed by the European Social Fund.

References

1. Allen, M.B., Behic, G.A., Trangenstein, J.A.: Multiphase Flow in Porous Media. Lecture Notes in Engineering, Springer, New York (1988)
2. Barrow, H., Rao, T.L.S.: The effect of variable beta on free convection. Br. Chem. Eng. **16**, 704–709 (1971)
3. Cheng, P., Minkowycz, W.J.: Free convection about a vertical flat plate embedded in a porous medium with application to heat transfer from a dike. J. Geophys. Res. **82**, 2040–2044 (1977)
4. Cheng, P., Chang, I.-D.: Buoyancy induced flows in a saturated porous medium adjacent to impermeable horizontal surface. Int. J. Heat Mass Transf. **19**, 1267–1272 (1976)
5. Furumoto, A.S.: A systematic program for geothermal exploration on the island of Hawaii. Paper presented at the 45th Annual International Meeting, Society of Exploration Geophysicists, Denver, CO, 12–16 Oct. 1975
6. Ingham, D.B., Pop, I. (eds): Transport Phenomena in Porous Media. Pergamon, Oxford (1998), II (2002), III (2005)
7. Ingham, D.B., Bejan, A., Mamut, E., Pop, I.: Emerging Technologies and Techniques in Porous Media, NATO Science Series II. Mathematics, Physics and Chemistry, vol. 134. Springer, Netherlands (2004). doi:10.1007/978-94-007-0971-3
8. Lai, F.C.: Mixed convection in saturated porous media. In: Vafai, K. (ed.) Handbook of Porous Media, pp. 605–661. Marcel Dekker, New York (2000)
9. Nazar, R., Arifin, N.M., Pop, I.: Free convection boundary layer flow over vertical and horizontal flat plates embedded in a porous medium under mixed thermal boundary conditions. Int. Comm. Heat Mass Transf. **33**, 87–93 (2006)
10. Nield, D.A., Bejan, A.: Convection in Porous Media, 2nd edn. Springer, New York (1999)
11. Lesnic, D., Ingham, D.B., Pop, I.: Free convection boundary-layer flow along a vertical surface in a porous medium with Newtonian heating. Int. J. Heat Mass Transf. **42**, 2621–2627 (1999)
12. Pop, I., Ingham, D.B.: Convective Heat Transfer: Mathematical and Computational Modelling of Viscous Fluids and Porous Media, pp. 381–430. Pergamon, Oxford (2001)
13. Rees, D.A.S., Pop, I.: Free convection induced by a vertical wavy surface with uniform heat flux in a porous medium. J. Heat Transf. **117**, 547–550 (1995)

14. Vajravelu, K., Sastri, K.S.: Fully developed laminar free convection flow between two parallel vertical walls, I. Int. J. Heat Mass Transf. **20**, 655–660 (1977)
15. Vajravelu, K., Cannon, J.R., Leto, J., Semmoum, R., Nathan, S., Draper, M., Hammock, D.: Nonlinear convection at a porous flat plate with application to heat transfer from a dike. J. Math. Anal. Appl. **277**, 609–623 (2003)
16. Vajravelu, K., Prasad, K.V., Van Gorder, R.A., Lee, J.: Free convection boundary layer flow past a vertical surface in a porous medium with temperature-dependent properties. Transp. Porous Med. **90**, 977–992 (2011). doi:10.1007/s11242-011-9827-5

On Some Functions of the MES Applications Supporting Production Operations Management

Péter Bikfalvi , Ferenc Erdélyi, Gyula Kulcsár, Tibor Tóth and Mónika Kulcsárné Forrai

Abstract Integrated computer applications play an increasingly important role in the planning and control of production systems and processes. The model-based decision support functions of business and manufacturing processes can be classified into different hierarchical levels according to their functions, objects and time horizons. Manufacturing Execution System (MES) applications represent the most challenging part of planning and control activities in Manufacturing Operations Management (MOM) in supporting its Production Operations Management (POM) activities. The ISA-95 standard delivers useful methodology and several models with the most important functions to be embedded in the shop floor level control loops. Two methods for improving the quality of POM are presented: a proactive one, using simulation-based fine scheduling and a reactive one, based on evaluation of some Key Performance Indicators (KPIs) determined from the efficiency analysis of shop-floor production data. Both methods exploit advantages of software applications used in different MES components.

1 Introduction

Nowadays, the need for global competitiveness and for sustainable development stimulates enterprises to integrate their profit expectations and quality requirements with flexible adaptation to market requirements when they are planning and carrying out their comprehensive business strategy. This is especially true when considering discrete manufacturing companies, where the standard activities of Business Management (BM) can be modelled as a multi-level matrix-like structure. This is characteristic to a great extent both of small and medium-sized enterprises, as well as for large multi-national companies (Fig. 1).

P. Bikfalvi(✉) · F. Erdélyi · G. Kulcsár · T. Tóth · M. K. Forrai
Department of Information Engineering, University of Miskolc,
Miskolc Egyetemváros 3515, Hungary
e-mail: bikfalvi@uni-miskolc.hu

G. Bognár and T. Tóth (eds.), *Applied Information Science, Engineering and Technology*, Topics in Intelligent Engineering and Informatics 7, DOI: 10.1007/978-3-319-01919-2_7,

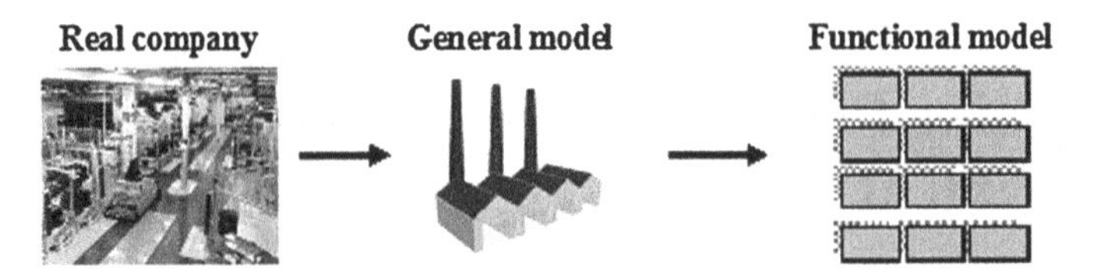

Fig. 1 Creating an abstract functional model of a company

Effective BM decisions can only be achieved by using integrated Information and Communication Technologies (ICT) solutions. Computer integrated application systems with well-adjusted models and tailor-made databases are expanding dramatically. However, total integration and automation is far from being achieved, and the "unmanned factory" still seems to remain a future goal.

Enterprise Resources Planning (ERP), Product Lifecycle Management (PLM), Supply Chain Management (SCM), Customer Relationship Management (CRM), Manufacturing Resources Planning (MRP II), Manufacturing Execution Systems (MES) and Shop-floor Control Systems (SCS) are known software components of an integrated computer application system that may serve BM goals. These software components use multi-layered data models, thousands of data objects and an even greater number of attributes. On the other hand, computer applications used in planning and control activities of BM must be integrated with engineering, marketing, finance, HRM (Human Resource Management), logistics and even other activities. That is, they are very complex. Moreover, components of software packages are developed independently by different vendors. Integration faces serious problems in this respect, too [20].

These are the main reasons why it is such hard work to build up a coherent, computer application system to support all the activities of a hierarchical BM from the strategic level down to the level of production control [29].

It is well known that the effective control of complex and complicated processes requires hierarchical architecture and feedback control loops. Feedback control calls for the measuring of specific quantities (based on physical processes) at the production execution level and needs aggregated, averaged, cumulated or more general derived indicators to measure the abstract "performance" of production processes at different managerial levels. Industrial experience shows that these indicators, also called Key Performance Indicators (KPIs), can play an increasingly significant role in assuring stable and successful production and in improving the efficiency of production both at the local (shop floor) or global (enterprise) level [1, 7, 28].

In the early days of Computer Integrated Manufacturing (CIM), developers thought that integration of functional software components means only common database and communication network protocol issues. However it turned out later that successful integration demands a common semantic referential approach, common concepts and methods in the applied models.

In the last two decades, different ERP, SCM and CRM applications on the business level, CAD, PLM, and CAPP applications on the engineering design or planning level and last but not least PAC (Production Activity Control) applications (such as CNC, PLC, Robot Control (ROC), Measuring Device Control (MDC), DCS, SCADA, etc.) on the process execution level have been implemented for improving production effectiveness and competitiveness. Experience in practise has confirmed the existence of a gap between the business-engineering levels and the PAC level. To eliminate this gap, experts (in production management, in IT, and in control systems) expressed the need for designing a new application system to support manufacturing operations management functions.

The general collective name of this software package was first called the MES—Manufacturing Execution System.

2 Literature Review

2.1 The Functional Model of Production Systems

There are several approaches concerning the matrix-like functional modelling of production systems. The hierarchical and horizontal structure of the company determines the matrix dimensions. The in-company information system supported by these (computer integrated) applications, as well as the control activities, more or less follow this structure.

The integrated software systems used in management of discrete manufacturing processes can be classified into four hierarchical groups according to the fields and time horizons of the different supported activities, as follows [8]:

1. Enterprise Resources Planning (ERP)—at the enterprise or business level,
2. Computer Aided Production Engineering (CAPE)—at the engineering level,
3. Manufacturing Execution Systems (MES)—at the production operation level,
4. Production Activity Control (PAC)—at the process execution level.

Figure 2 presents some of the software components of each group, though it is far from presenting all of the possible options available on the market. Each software object uses specific models for planning and control decisions at each level of the hierarchy. Each model presents state variables, model parameters, goal functions, and constraints. Solving and optimising planning problems on the higher levels of hierarchy results in specific goal parameters that extend the set of constraints of the model problems at the lower level [3].

It is worth noting here that the allocated time for decision making decreases as the level of hierarchy becomes lower.

Figure 2 shows that MES applications are connected with other applications of the same level, but also with applications of the upper (engineering and business) levels, as well as those of the lower (execution) level. Modelling and functionality aspects of the MES applications, as well as their connection to the upper levels, are supported by the ISA-95 standard [8, 10, 27].

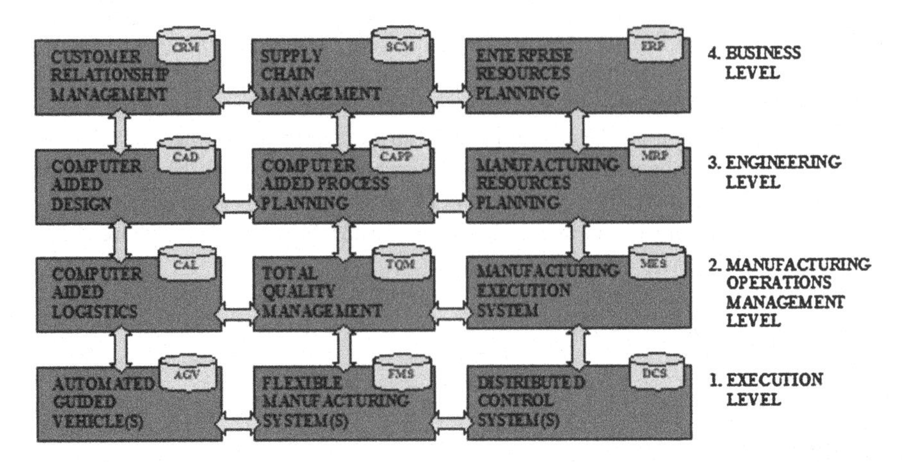

Fig. 2 Typical software components of a functional model

2.2 Relationship Between the MES Components and the ISA-95 Standard

The need for computer support at the operative production management level came up many years ago. The international non-profit organization MESA International (Manufacturing Enterprise Solutions Association) was founded in 1991 in order to exchange knowledge, experience, and best practices on MES applications among manufacturing experts, system integrators and software developers.

The term MES arose at the same time and stood for *manufacturing execution systems*. In 2004 MESA changed the term to *manufacturing enterprise solutions*, because they considered that MES must be more than just a system for production control [26]. They considered that issues such as machine maintenance, quality control, inventory control, product data and product life-cycle management needed to belong to the MES domain, together with data acquisition and analysis for production assessment, as well as compiling different types of documentation. However, today one may conclude that these goals were very ambitious; despite the fact that progress had been made towards meeting them, in practice there is still a lot of work to do in completing the integration of all these functions. The present chapter is restricted to analysing only the main planning and control functions to be implemented in the MES application systems and how these functions are supported by the ISA-95 standard.

ISA (the International Society of Automation) was founded in 1945 as a non-profit organization to support standardisation in the field of industrial automation. Some well-known ISA standards are: ISA-84 (Functional Safety), ISA-88 (Batch Control), ISA-99 (Manufacturing and Control Systems Security) [25].

The ISA-95 standard is dedicated to Enterprise-Control System Integration. The standard describes the methodology, the way of working, thinking and communicating

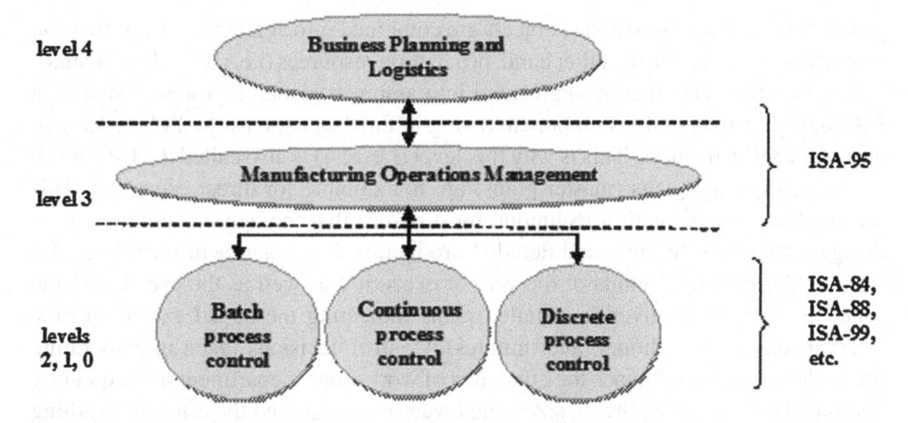

Fig. 3 The ISA-95 functional hierarchical control model [26, 27]

when analysing a manufacturing company aiming at ERP-MES integration. The documents contain models focusing on specific aspects of this integration, serving the main goal of the standard: full integration of production and business processes. The integration is presented as a hierarchical, multi-level model (Fig. 3). The upper level (Level 4) of this model represents Business Planning and Logistics (BPL), while at the lowest level (Levels 2, 1 and 0) the supervisory and direct control activities of manufacturing processes are performed.

ISA-95 focuses on Manufacturing Operation Management (MOM) situated at Level 3 of the model, and its integration with the upper BPL level. It is worth noting here that integration between the two levels is necessary because several functions like production scheduling, inventory control, quality assurance or maintenance straddle the border between the two levels. On the other hand, one may distinguish clearly the activities that belong to the enterprise (business) management domain (Level 4) from those belonging to the production management domain (Level 3 and lower).

2.3 The Role of MES

Today the dynamic and structural complexity of manufacturing systems makes traditional production planning and control (PPC) activities increasingly difficult, and the methods and models used are insufficient to face the dynamic changes of the global market. The cumulative nature of production processes in time, the increasing intensity of production processes as well as the importance of customer demands and deadlines emphasise the outstanding role of PPC, and, in this respect also, the role of integrated computerised solutions [11].

At the BPL level, production planning and scheduling (PPS) covers longer time periods (weeks, months), and it results in aggregated production plans (master plans) that are delivered to the production departments. Aggregation means that individual production (distinct but similar) objects are combined into aggregate groups that can be planned together. On the other hand, production resources (i.e. particular machines or labour pools) can also be aggregated into assumed resource groups. Usually an ERP application system's component acting at this level solves the PPS tasks with acceptable effectiveness. This is why this level (Level 4) is also called the ERP level.

The aggregated plans (master plans) are not suitable for direct execution. This means that the aggregation technique must assure that the aggregate plan can be disaggregated into feasible and detailed production (i.e. manufacturing) schedules whenever necessary. Details of the way of execution as well as the execution time schedule have to be given in detailed plans indicating the actual execution time interval (days, shifts, hours, and minutes). Control decisions taken in elaborating these plans directly influence the efficiency of workshops (departments), the quantity and quality of products, the lot sizes, the inventory levels, and the efficient handling of materials and utilisation of resources. POM acts at Level 3 and has to solve all these tasks. As MES application systems act at this level (and also have to solve the PPC tasks effectively), this level (Level 3) is also called the MES level.

Levels 2, 1 and 0 of control belong to the execution level, where the production (manufacturing) processes take place. Time intervals of minutes, seconds or often below are characteristic. Here, control functions are performed by SCADA and/or DCS application systems, process controllers, PLCs, intelligent sensors and actuators, or other computerized systems (e.g. CNC, robot controllers, navigation systems, identification systems, cell controllers, measuring devices, etc.) serving automation.

Whatever the upper hierarchical level is, PPS typically runs according to the rolling time horizon principle. That is, a candidate plan is created for the actual time horizon and decisions of the first few time-periods of this horizon are executed accordingly. Then, the candidate plan is revised and re-planned, if necessary. Revision/re-planning is due to the uncertainties and unpredicted events that may occur in the demand list and—often—in the production (manufacturing) processes.

According to the rolling time horizon, PPC application systems cover two main stages or hierarchical levels: master planning and master scheduling are realised in the first stage (on the ERP level), while fine or detailed scheduling and execution control is in the second stage (on the MES level).

The main goal of the first stage is to balance the capacity and customer demands. The PPC system includes the scheduling function that produces master production plans for material and capacity requirements.

The second stage deals with sequencing of production orders (jobs) that have already been released for production in the actual time frame. It also decides exactly when and on which machines or workplaces the dependent jobs should be executed. The main goals of this second stage are to avoid tardiness of jobs, to minimise flow times of parts and products, and to maximise utilisation rates of machines/workplaces. For solving such detailed or fine scheduling tasks, advanced scheduling models and methods are needed [13].

Corrections of the production plans must be performed as effectively as possible (as close as possible in terms of both time and location). The control role of MES applications is outstanding in this respect: here, by processing the abstract information with the modelled real objects and processes, production planning and control can be directly and efficiently optimised, as production performance can be improved as well.

2.4 Relationship Between MES and MOM

ISA-95 presents a generic activity model of MOM (Fig. 4), with four main classes of information exchange between POM and BPL, respectively between MES and ERP applications. They are as follows:

1. Operations definitions,
2. Operations (resources) capabilities,
3. Operations (and their intensities) requests,
4. Operations response (operations results and their performance).

The ISA-95 generic model summarises all operative activities to be carried out at MOM level [9]. The model is general, so it could be applied not only for the manufacturing departments (plants, workshops) but other departments as well where operative production activities may take place, such as maintenance, quality testing or material handling and storing (Fig. 5).

Therefore, as shown in Fig. 5, tasks and activities of the operative MOM refer to four main functional areas of operation company. They are as follows:

1. Production operations management,
2. Maintenance operations management,
3. Quality test operations management,
4. Inventory operations management.

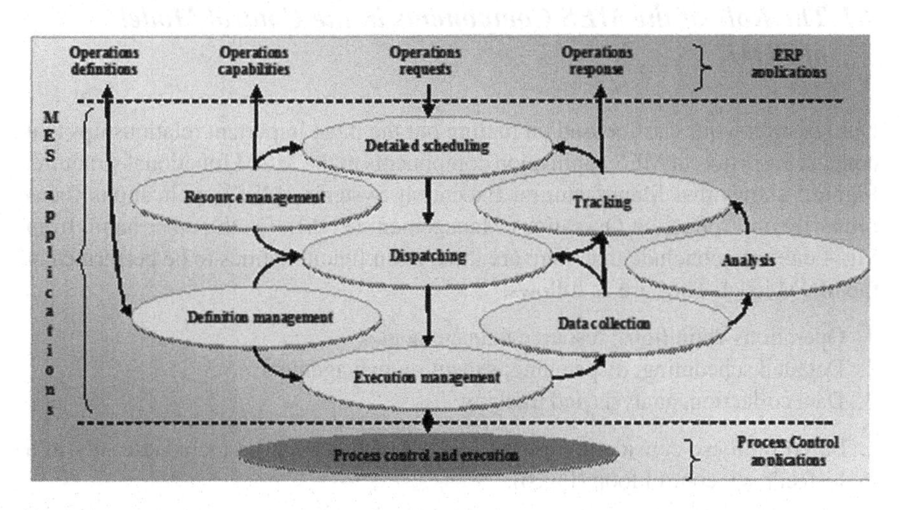

Fig. 4 ISA-95 generic activity model of MOM [26]

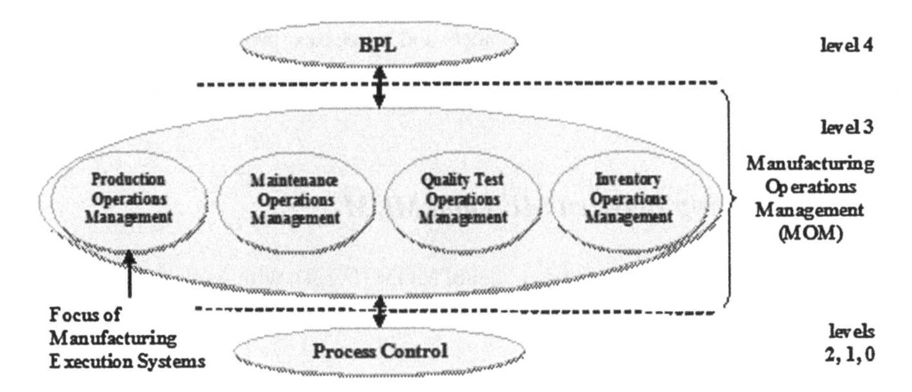

Fig. 5 Relationship between MES and MOM [26]

One may also notice from Fig. 5 that the main focus of MES applications is on Production Operations Management (POM); however, strong integration with other components of MOM level areas is indispensable in the everyday life of a company.

Since the ISA-95 standard was published, the international literature has presented a number of refereed publications that confirm the fact that the standard-defined terms, models and methods contribute to a large extent to the effectiveness of workshop level analysis for production management, as well as to the development of decision support functions implemented in the MES applications [25–27].

3 Improving Planning and Control Activities at MOM Level

3.1 The Role of the MES Components in the Control Model of POM

Our research work was focused on finding out the most important relationships that connect the different MES application components to the MOM functional structure. Figure 5 shows that Manufacturing Execution Systems (MES) applications focus primarily on Production Operations Management (POM). On the other hand, from Fig. 4 one may conclude that there are three main functionalities to be performed at the MOM level. They are as follows:

1. Operations definitions, resource management,
2. Detailed scheduling, dispatching, execution management,
3. Data collection, analysis and tracking.

Based on these considerations, the above functions can be included easily in a multi-feedback control loop (Fig. 6).

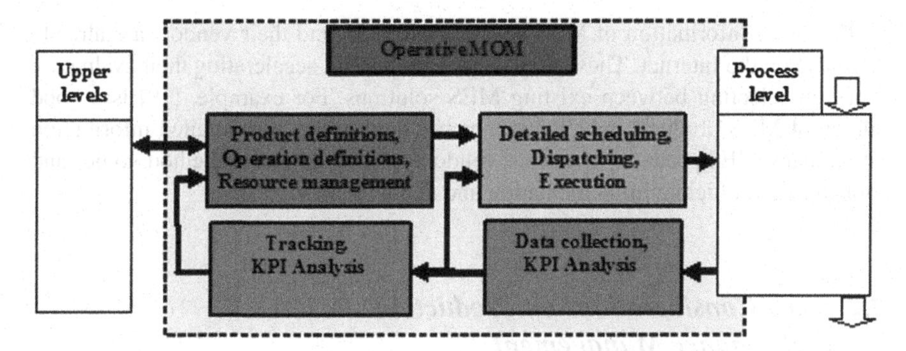

Fig. 6 Feedback control system of operative MOM

From the ICT point of view, operative MOM activities have to be incorporated in the MES application modules. These software components are "horizontally" connected with each other, but they are "open" to "vertical" information exchange with the applications of the upper hierarchical level (CAD/CAM, ERP, SCM, etc.) as well as with those from the lower level (SCADA, DCS, etc.).

Industrial experience confirms that over the past years personnel involved in the everyday operative MOM of production departments have developed their own computer aided (usually Excel or Access based) "solutions". These partial and specific applications are attuned to their user needs, and they contain many subjective elements. Unfortunately, they are usually poorly documented and they do not meet the modern integration requirements. Moreover, all of the knowledge is in hands of just a few people, which puts the plant's operation and control continuity at high risk.

The situation is somewhat similar concerning products offered by different software companies. Typically, these products support only some of the presented activity model objects and do not assist all spectra of POM functions. Some vendors offer advanced reporting tools (historians), but do not provide scheduling functionality. Conversely, there are specialised Advanced Planning and Scheduling (APS) software components that do not provide any data reporting and/or analysis. Moreover, issues related to operations definitions or resource management are usually manually solved.

It is also true that due to the less general fields of activities as well as due to the burden of problems coming from local specificities, the full integration of MES applications still has a long and bumpy way to go. Cost-efficient interfaces, management of master data and software quality are issues to be considered [26].

Most commercially available MES applications offer their functionality in modules. Unfortunately, these modules usually do not coincide with those of the ISA-95 generic activities. The most important modules implementing parts of MES functionality are concerned with detailed production scheduling, product definition management and production execution management, history reporting (usually including an interface with Level 2 systems), and tracking, dashboards, workflow management, interface modules for ERP integration, plant modelling and simulation tools, etc.

Extensive information of MES software products and their vendors are already available on the Internet. These may help companies in accelerating their evaluation work on selecting between existing MES solutions. For example, [19] is a good survey of MES products, which contains not only useful and detailed information about many MES products of various vendors, but includes also methodologies and tools for their efficient implementation and deployment.

3.2 Some Considerations on Production Performance Management

Once a company has customer orders (demands) and schedules of available resources, measuring the performance of production is a very important issue, from the points of view of both production planning and production control. However, developing appropriate production models with different complexity makes it possible to analyse, classify and manage production performance issues.

Both ERP and MES systems use various mathematical approaches to model production processes. Most of these models consist of different production equations representing the dynamic state equilibrium of production processes. Discussions on these production equations demonstrate that three KPIs of great importance can be defined, the determination of which in the decision domain of PPC is a necessary and at the same time a satisfactory condition to quasi-optimal realisation of a production goal defined in a well-established way and meeting all the significant constraints. The relevant mathematical model was named as "production triangle" model [4, 11] and the three essential KPIs are:

1. Delivery capability (on ERP level) or throughput time of internal orders (on MES level),
2. Inventory (on ERP level) or work-in-process (WIP) (on MES level),
3. Return of assets (on ERP level) or utilisation of resources (on MES level).

Based on theoretical considerations as well as on experience in PPC, it can be inferred that improvement of any of these three performance indices is accompanied by deterioration in the other two; therefore the three indices must be managed only together. The outlined optimisation problem logically extends to all the details of production, including especially manufacturing. It also means that for optimisation of production planning and control tasks, examination of the three KPIs is not only a necessary but at the same time a satisfactory condition. It can be also conceived that the method can be extended logically to any manufacturing enterprise and engineering industry operating on the basis of discrete manufacturing.

It is worth mentioning here that other performance indicators could be defined, as well. Such indices could be: the number of finished-products produced over a time period, quality of the products in question, the unit cost of a product or its components, the net profit producing ability of production, the completion time of orders (especially lateness of jobs over deadline) [12, 29].

All of the above-mentioned indicators are actually very important. Evaluating them is an everyday duty of production management. However, it can be shown that several of them can be expressed or influenced directly by the considered three main KPIs, while the others are strictly influenced only by economic and sales decisions and conditions, and therefore influence the production process in only an indirect way.

Taking into consideration the profit-oriented behaviour of enterprises, production planning tasks have to be solved to meet the following requirements:

1. delivery capability/throughput time should be according to the customer's order in every case (in general, as short a term of delivery or throughput time as possible);
2. stock/WIP level should be as low as possible (valid for all raw materials, semi-finished products, finished products/articles, spare parts, etc.);
3. utilisation of resources/return of assets (machines, tools, workforce, etc.) should be within the limits prescribed by the management goals.

At the ERP level, when adapting these tasks to a given enterprise one usually gets a large-sized optimisation problem. From a mathematical modelling point of view it can be recognised that the three KPIs—namely delivery capability, stock level and utilisation of resources—are functions of the BM decision parameters of "internal orders". The KPIs partially depend on how and to what extent the external order book changes. In addition, they are partially dependent on the method of scheduling and on all the constraints of production, as well as on the factors related to the specific technological processes employed. That is, at the ERP level, the basic task of the PPC system is to determine the manufacturing and the purchasing orders (internal orders) taking into consideration the constraints of the current production environment in such a way that the given enterprise should operate near to the "optimal working point" from the aspects of the three KPIs (Fig. 7). In general, the optimisation

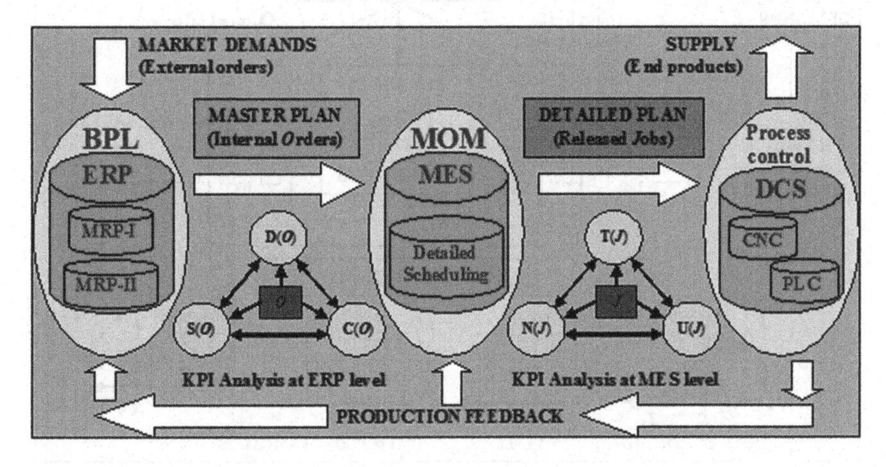

Fig. 7 Evaluation of production objectives based on KPI analysis

problem is difficult to solve; heuristics or suitable procedures based on some kind of Operations Research method are needed.

Evaluation and fine-tuning of a PPC system at the ERP level is usually made during its installation and first runs, assuring its usability and applicability for years. However, small corrections or compensations may later occur every now and again.

At the MES level, the three KPIs—namely throughput time, WIP level and utilisation of machines/workplaces—are functions of the MOM decision parameters of released jobs and operations. The KPIs partially depend on how and to what extent the internal order list changes, but are most strongly influenced by the availability of resources, as well as by any unpredicted events that may occur. So at the MES level, the basic task of the PPC system is to sequence the jobs and operations in a detailed manner by taking into consideration the actual constraints of the current production environment. Here, evaluation and fine-tuning of the PPC system requires usually continuous and interactive human supervisory control and intervention.

As an example, the most important shop-floor management goals (KPIs) can be followed and deduced from the model represented in Fig. 8.

The long-term business goal of a manufacturing company is to maximise its profit. However, this goal can only be achieved by a series of some other production goals to be realised in shorter time periods. These short-term goals are connected with the output characteristics of production (manufacturing). The up-to-date MES, by means of production monitoring and data acquisition, observes the formation of the "production triangle" KPIs through the comparison of the planned and realised values. Hence, the KPIs have an important role not only in the optimisation tasks of proactive production planning period, but also by acting as feedback signals of the production control loop in the execution phase. Therefore, it is very important for

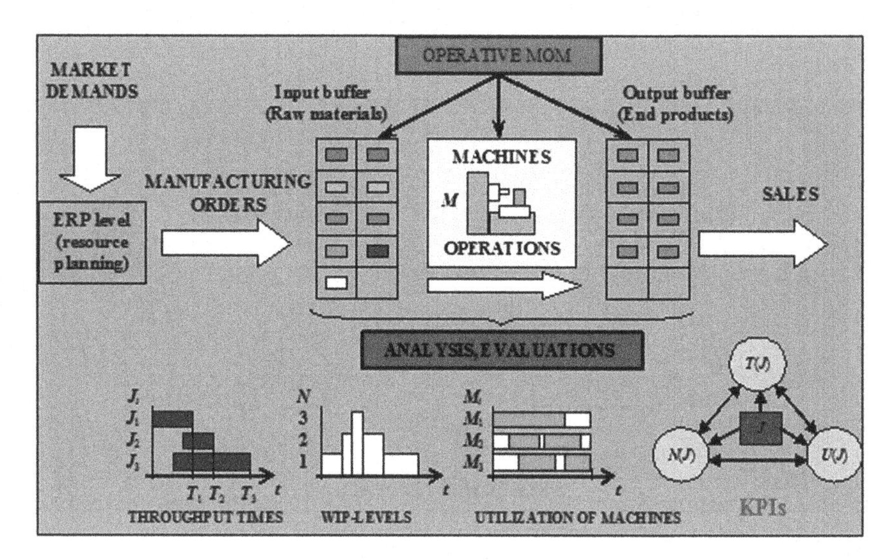

Fig. 8 Shop-floor performance evaluation based on KPI analysis

production planners and dispatchers to be fully aware of the capabilities and limits of their plant (workshop), and to recognise potential sources of conflict that appear in the course of measuring their production performance.

3.3 Production Planning and Scheduling Issues

3.3.1 Basic Scheduling Approaches

MOM has to perform the real-time supervision and control of manufacturing operations. The most important tasks to be performed are defining and fine scheduling of production operations, their dispatching, and managing the execution processes.

Among these activities, the detailed scheduling of projects, tasks, jobs and operations (prescribed by the hierarchically higher ERP and CAE level) represents one of the key issues to be efficiently and effectively realised. Detailed scheduling must take into consideration the related technological plans, the availability of resources (machines, people, raw materials, etc.) and must analyse the actual restrictions and options. Then it has to connect the operations with the corresponding resources, and it has to schedule their launching sequence and time [7, 8]. That is, fine scheduling is the most important decision support tool at POM level to be implemented as a MES software component.

Decisions in scheduling can basically be made in a predictive or a reactive mode. In practice, rolling-time horizon based predictive scheduling results in a detailed execution program for the actual time-period, which, hopefully, is executed correspondingly. However, one may observe an increasing demand for fast handling of uncertainties and unpredicted events that may happen during the execution of these programs. In order to face such problems, suitable reactive scheduling procedure must be used. Based on the different events that occur and on some kind of logic, the reactive scheduling must select the best decision for the most adequate intervention activity. However, the decision procedure is far from being always sound and clear.

Discrete manufacturing processes present a large variety of diverse and very distinct technologies that claim specific models when designing their efficient control systems. In this respect, there is a special demand for accurate modelling, formulation and solving of scheduling problems.

3.3.2 Scheduling Methods

Scheduling is in fact an optimisation problem. There is a huge amount of literature that deals with such problems. Their complete review and classification exceeds the scope of this chapter. Therefore, only some of the most important mathematical approaches are shortly enumerated below:

- Mathematical Programming: Formulations and Applications (e.g. Linear, Nonlinear, or Integer Programming, Set Partitioning, Covering and Packing, Disjunctive Programming, etc.),
- Exact Optimisation Methods (e.g. Dynamic Programming, Constraint Programming, Branch-and-Bound methods, etc.),
- Heuristic and Meta-heuristic Methods (e.g. Basic and Composite Dispatching Rules, Beam Search, Local Search, i.e. Simulated Annealing, Tabu Search, Genetic Algorithms, etc.).

3.3.3 Variety of Scheduling Problems

The most difficult problem that the MES application component of fine scheduling faces during its application is related to the great variety of manufacturing process models [2]. That is, flexibility and local tuning have to be also implemented.

Solution of a scheduling problem must deal with at least three important issues. These are as follows:

1. Resource environment: this takes into consideration all characteristics of the resources concerned (machines, workplaces, workers, raw materials, etc.) and of the relations among them, with special attention paid to the features of the jobs operations to be executed;
2. Job characteristics: rules, restrictions, execution features and alternatives;
3. Production policy: priorities, rules, objective functions, KPIs.

The simplest scheduling problem is represented by the one-machine type model, which refers to jobs containing one operation to be performed by only one machine. In the case of parallel machine models, the one-operation containing jobs are executed simultaneously on different machines/workplaces. Depending on the machines' simultaneous working capability, three further model types are distinguished [23]:

- Identical parallel machines model, where all machines perform identically (for the processing time τ_{ij} of job J_i on machine M_j we have $\tau_{ij} = \tau_i$ for all machines);
- Uniform parallel machines model, where the operations intensity on each machine varies according to that machine ($\tau_{ij} = \tau_i / s_j$ where s_j is the speed of machine M_j);
- Unrelated parallel machines model, where the operations intensity on each machine varies according to the job (operation) performed on that machine ($\tau_{ij} = \tau_i / s_{ij}$ for job-dependent speeds s_{ij} of machines M_j).

The shop-floor models involve more machines and more jobs containing more than one operation. However, each operation can be performed on a given machine. According to different prescriptions, from this general shop specific models can be formed, as follows [5]:

- Job Shop Model: the number of operations is job-dependent and they have special precedence relations;
- Flow Shop Model: a special case of the job shop model in which the number and sequence of operations is fixed for any job;
- Open Shop Model: a general shop model in which there are no precedence relations between the operations;
- Mixed Shop Model: a combination of the above models;
- Flexible Shop Model: a further extension of the general model, where flexibility refers to the scheduling possibilities.

In the above enumerated models, each operation is allocated to a given machine (dedicated machine), while in the case of the flexible models (Flexible Flow Shop or Flexible Job Shop) a given operation can be performed on any of the machines of a specified machine group. In this way, the scheduling problem is extended with selection of a specific machine. It is worth to note that the machines from the selected group can simultaneously work identically or uniformly or even unrelated [24, 30].

The Extended Shop Models represent a new generation of the scheduling problem class (Extended Flexible Flow Shop, Extended Flexible Job Shop, etc.). It often happens in the manufacturing systems that there are some resource objects (machines/workplaces) that can perform more than one technological operation. In this case, some operations can be grouped in bigger units such as technological steps or even execution steps, which could be considered operation units for scheduling [14, 15].

4 Performance Analysis at MES Level

4.1 Some Fundamental Concepts of Production Systems

At the shop-floor level, there are three important rates (intensities), which represent basic information received from the ERP level by the MES level. They are as follows:

1. The arrival rate of the internal orders (jobs);
2. The demand rate given for the jobs. (This is the quotient of the ordered quantity, and the duration of the production time window);
3. The production rate, which is the inverse of the operation time.

At the workplace level, the production rate is a parameter of the technology process plan. The planned production rate influences the operation time, the tool utilisation and the product quality, too. In the case of a bottleneck, especially in the case of multi-stage production routing, the local production rate greatly influences the characteristics of the realisable schedules.

In a flow-shop manufacturing system (Fig. 8) jobs arrive at discrete time instants T_{Ai}, for $i = 1, 2, \ldots, n$, while the finished jobs leave the system at T_{Ci}, $i = 1, 2, \ldots, n$, discrete time moments, too. If $A(t) = \sup_i (T_{Ai} \leq t)$ and $C(t) =$

$\sup_i (T_{Ci} \leq t)$ represent the number of jobs arriving respectively, leaving the shop in the time interval $0 \longrightarrow t$ and $N(t)$ is the number of jobs in progress (or WIP) in the system at the time moment t, the fundamental cumulative state equation of production will result as follows:

$$N(t) = N(0) + A(t) - C(t). \tag{1}$$

An important performance index of production can be pointed out as the average value $\bar{N}$ of the number of jobs in the system, related to the time interval $0 \longrightarrow t$, when considering a long-term run (theoretically $t \longrightarrow \infty$) [6]:

$$\bar{N} = \lim_{t \to \infty} \frac{1}{t} \int_0^t N(\tau)\, d\tau. \tag{2}$$

The stock level of production is a function of this averaged value $\bar{N}$, usually named the WIP level or Average Work Level (AWL).

The job cycle-time (flow time) of every job separately can be expressed as:

$$T_i = T_{Ci} - T_{Ai}. \tag{3}$$

Based on this, the average flow time (AFT) $\bar{T}$ value, as another important long-term performance index, can be conceived as:

$$\bar{T} = \lim_{k \to \infty} \frac{1}{k} \sum_{i=1}^{k} T_i. \tag{4}$$

In long-term performance analysis, if the jobs are independent of each other, $\bar{T}$ represents the expected value of planned lead-time. This index essentially influences the character of delivery capability, too. However, delivery capability can be measured in many other ways as well, like make-span (T_{ms}), number of late jobs (n_L), etc. For example, if every job has a corresponding due date, then delivery capability can also be measured by the number of jobs with lateness or the cumulated time of lateness.

Production managers consider these performance indices to be useful, but the most important performance index for the delivery capability is the customer service level index (I_{CSL}). Noting with n the number of released jobs, this index is defined as:

$$I_{CSL} = (n - n_L)/n. \tag{5}$$

Another essential long-time rate-type performance index is the average arrival rate of jobs $\bar{a}$, defined as:

$$\bar{a} = \lim_{t \to \infty} \frac{A(t)}{t}. \tag{6}$$

The most important result of long-time performance analysis is that the performance indices defined in Eqs. (2), (4) and (6) are connected with each other in the following way:

$$\bar{N} = \bar{a} \cdot \bar{T}. \tag{7}$$

This clear interdependence of the three main KPIs for the “one or m-machine job-shop” abstract system model is the simplest “production equation” known as Little’s Law, firstly presented in 1961 [18]. The above equation is of great importance since it expresses a direct connection among the variables characterising the average state of production under general conditions. It is easy to observe that the connection under the above-defined conditions is linear, i.e. $\bar{N} = f(\bar{T})$ is a simple linear function. It is also important to note that Little’s Law is valid for both cases of deterministic or periodic arrival distribution in time and of even optional arrival distribution as well. Details are presented in [6, 21, 22].

4.2 Production Stability Issues

A fundamental requirement of manufacturing control policy is the stability of production. A production process is stable if the number of jobs in the system remains finite for an optional long time interval (i.e. they do not cumulate). In this case, buffer dimensions are feasible and the orders, sooner or later, will be certainly performed. This means that:

$$\overline{N} \leq N^*, \tag{8}$$

where the reference number N^* has a practical finite value. This condition can be performed if and only if the average value of technology operations times to be realised satisfies a special constraint, that is:

$$\overline{a} \leq \overline{q}, \tag{9}$$

where $\overline{q}$ is the average operation rate defined as:

$$\bar{q} = 1/\bar{\tau}, \tag{10}$$

and

$$\bar{\tau} = \frac{1}{n \cdot m} \sum_{i=1}^{n} \sum_{j=1}^{m} \tau_{ij} = \frac{\tau_L}{m \cdot n}, \tag{11}$$

where m is the number of machines (workplaces, resources) used, τ_{ij} is the technology operation time of job i on machine j and τ_L is the actual technological "load". Equation (9) expresses the trivial, but fundamental assumption that the content of intermediate buffers between machines (workplaces) cannot remain finite if the jobs arrive at a rate greater than their production rate (processing rate). On the other hand, for a finite T_s time horizon, one may write: $\bar{a} = n/T_s$ therefore $\overline{a} \leq \overline{q}$ which means:

$$\frac{n}{T_s} \leq \frac{n \cdot m}{\sum\limits_{i=1}^{n} \sum\limits_{j=1}^{m} \tau_{ij}} \quad (i.e.\ \tau_L \leq mT_s).$$

That is, in long term run, the load must be less then the capacity.

Another important rate-type performance index to be considered is the average utilization of resources, denoted by $\bar{u}$. This is in fact the rate of the time spent in work and the total time available. It was proved in [6] that for the j-th machine M_j results:

$$\bar{q}_j = \frac{n}{\sum\limits_{i=1}^{n} \tau_{ij}} \quad \text{and} \quad \bar{u}_j = \lim_{t \to \infty} \frac{1}{t} \sum_{i=1}^{A(t)} \tau_{ij} = \frac{\bar{a}}{\bar{q}_j}. \tag{12}$$

It is obvious that the utilisation rate of resources cannot be greater than 1 (i.e. 100 %). In the case of stable production, the following condition is also valid:

$$\bar{u}_j = \bar{a}/\bar{q}_j \leq 1 \ \text{ for } \ j = 1, \ldots, m. \tag{13}$$

Relation (13) means that any of resources (machines, work-places) can be loaded to its capacity limits only. This is the basic capacity constraint in any detailed scheduling of manufacturing operations. If Eq. (13) is fulfilled, i.e. the production is stable, it results in:

$$\bar{a} = \bar{u} \cdot \bar{q}. \tag{14}$$

Using Eqs. (7) and (14) it can be written:

$$\bar{N} = \bar{q} \cdot \bar{u} \cdot \bar{T}. \tag{15}$$

This last important relationship can be named as the "production triangle" equation. It expresses that if the operation rates are given in the technology process plans, and the lot sizes and the due dates are given in the manufacturing orders, the three performance indices of the production triangle model are interconnected and they cannot be improved independently of each other. For instance, if the jobs have rigorous deadlines T_{Di} and the condition $T_{Ci} \leq T_{Di}$ must be performed for every job i, then machine utilisation and stock level cannot be improved at the same time. Similarly, if one wants to keep machine utilisation at a high level then stock level and flow time cannot be decreased at the same time [12, 14].

4.3 Performance Analysis for a One-Machine Manufacturing System

Predictive, model-based off-line scheduling for one-machine (OM) systems is the simplest case. Jobs arrive in groups, at $t = 0$. Usually, the objective functions to minimise are the average flow time or the average number of jobs in process. If jobs have deadlines, minimising the number of late jobs or the time difference from the deadline is applied.

In the case of n simultaneously released jobs (i.e. in a group) $T_{Ri} = 0$ for $i = 1, \ldots, n$ the make-span results as:

$$T_{ms}^{OM} = \max_i(T_i) = \sum_{i=1}^{n} \tau_i = \tau_L. \tag{16}$$

The average job number $\bar{N}$ in the system for the considered make-span is a time average, which can be calculated with the help of the "unit step"-like function $N_i(t) = 1(t - T_{Ai}) - 1(t - T_{Ci})$, which represents the actual increase/decrease in the number of jobs. Taking into consideration that $\int_0^{T_{ms}} N_i(\tau)d\tau = T_i$ and $N(t) = \sum_{i=1}^{n} N_i(t)$, it yields:

$$\bar{N} = \lim_{t \to T_{ms}} \frac{1}{t} \int_0^t N(\tau)d\tau = \frac{1}{T_{ms}} \int_0^{T_{ms}} \sum_{i=1}^{n} N_i(\tau)d\tau$$

$$= \frac{1}{T_{ms}} \sum_{i=1}^{n} \int_0^{T_{ms}} N_i(\tau)d\tau = \frac{\sum_{i=1}^{n} T_i}{T_{ms}}. \tag{17}$$

The utilization rate can be expressed as:

$$\bar{u}^{OM} = \frac{\sum_{i=1}^{n} \tau_i}{T_{ms}} = 1, \tag{18}$$

and it is unitary valued (100 %), since on the make-span time horizon there is no waiting time for the machine.

The average production rate can be also easily expressed as:

$$\bar{q}^{OM} = \frac{n}{\tau_L} = \frac{n}{\sum_{i=1}^{n} \tau_i}. \tag{19}$$

The average throughput time (flow time) of the jobs will be:

$$\bar{T}^{OM} = \frac{\sum_{i=1}^{n} T_i}{n}. \tag{20}$$

Replacing (18), (19) and (20) in (15) and comparing to (17) the following result reflecting clearly the above conclusions - is obtained:

$$\bar{N}^{OM} = \bar{q}^{OM} \cdot \bar{u}^{OM} \cdot \bar{T}^{OM} = \frac{n}{\sum_{i=1}^{n} \tau_i} \cdot \frac{\sum_{i=1}^{n} \tau_i}{T_{ms}} \cdot \frac{\sum_{i=1}^{n} T_i}{n} = \frac{\sum_{i=1}^{n} T_i}{T_{ms}} = \bar{N}^{OM}. \tag{21}$$

In conclusion, the production equation is verified and is valid.

4.4 Performance Analysis for m-Machines Flow-Line Manufacturing System

In the case of a flow-line type manufacturing system with m machines (Fig. 9) Little's Law holds true for each machine separately, that is:

$$\bar{N}_j = \bar{a}_j \, \bar{T}_j, \quad \text{for} \quad j = 0, 1, 2, \ldots, m, \tag{22}$$

where $\bar{N}_j$ is the number of jobs held on the j-th work place (in machine M_j + buffer), and $\bar{a}_j$ is the average arrival rate of jobs to M_j. The average number of jobs $\bar{N}^{FS}$ held on in the flow-line system (FS) is the sum of the average number of jobs held at different workplaces, because any job can be physically only at one machine/input buffer at one time moment. That is:

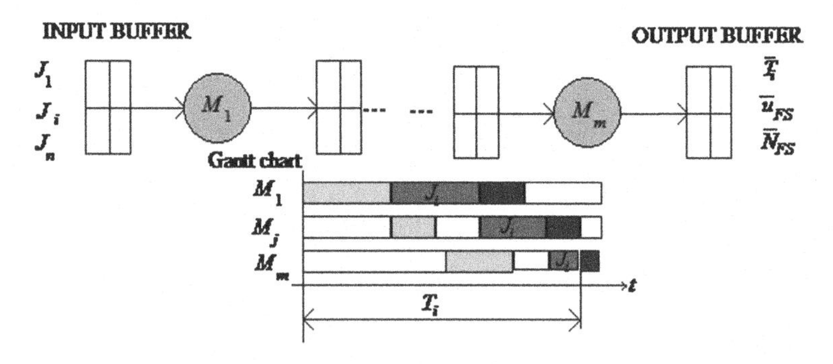

Fig. 9 Simple model of a flow-line type manufacturing system

$$\bar{N}^{FS} = \sum_{j=1}^{m} \bar{N}_j. \tag{23}$$

Similarly, the job cycle time (throughput time) T_i of the jobs is the sum of the times spent on the serially connected machines, since the jobs go through each machine in the same sequence and thus the manufacturing and waiting times are added. That is: $T_i = \sum_{j=1}^{m} T_{ij}$ and therefore:

$$\bar{T}^{FS} = \sum_{j=1}^{m} \bar{T}_j, \tag{24}$$

where $\bar{T}_j = \frac{\sum_{i=1}^{n} T_{ij}}{n}$ is the average flow time on machine M_j.

In long-term functioning, the average demand for operations on every machine will be equal to the average arrival rate, since in long-term runs every job has to pass every machine. This means that:

$$\bar{a}_j = \bar{a} = \lim_{t\to\infty} \frac{1}{t} \int_0^t A_J(\tau) d\tau. \tag{25}$$

From relations (22), (23), (24) and (25) it results that:

$$\bar{N}^{FS} = \bar{a}_j \cdot \bar{T}^{FS} = \bar{a} \cdot \bar{T}^{FS}. \tag{26}$$

Based on (12), the utilization rate of each machine is:

$$\bar{u}_j = \lim_{t\to\infty} \frac{1}{t} \sum_{i=1}^{A_J(t)} \tau_{ij} = \lim_{t\to\infty} \frac{1}{t} \int_0^t u_j(\tau) d\tau = \frac{\bar{a}_j}{\bar{q}_j} = \bar{a}_j \bar{\tau}_j, \tag{27}$$

where the average operation rate $\bar{q}_j$ of machine M_j for the $\{J_i\}$ released jobs was replaced according to Eq. (11).

The average utilization rate $\bar{u}^{FS}$ of the whole flow-line system will be:

$$\bar{u}^{FS} = \frac{1}{m} \sum_{j=1}^{m} \bar{u}_j = \frac{1}{m} \cdot \bar{a} \cdot \sum_{j=1}^{m} \bar{\tau}_j = \bar{a} \cdot \frac{1}{m} \cdot \sum_{j=1}^{m} \frac{1}{n} \cdot \sum_{i=1}^{n} \tau_{ij} = \bar{a} \cdot \frac{1}{n \cdot m} \cdot \sum_{j=1}^{m} \sum_{i=1}^{n} \tau_{ij}. \tag{28}$$

If $\bar{q}^{FS}$ notes the so-called characteristic intensity (average operating rate) of the flow-line manufacturing system, it is defined as:

$$\bar{q}^{FS} = \frac{1}{\bar{\tau}} = \frac{n \cdot m}{\sum_{j=1}^{m} \sum_{i=1}^{n} \tau_{ij}}, \tag{29}$$

which by replacing it in (28) yields:

$$\bar{u}^{FS} = \bar{a} \cdot \bar{\tau} = \frac{\bar{a}}{\bar{q}^{FS}}. \tag{30}$$

From the obtained relations, one can easily verify that the "production triangle" equation is valid for the long-term operation of flow-line manufacturing systems, too. Replacing (28) in (15) accordingly and using (25) the following result is obtained:

$$\bar{N}^{FS} = \bar{q}^{FS} \cdot \bar{u}^{FS} \cdot \bar{T}^{FS} = \bar{q}^{FS} \cdot \frac{\bar{a}}{\bar{q}^{FS}} \cdot \bar{T}^{FS} = \bar{a} \cdot \bar{T}^{FS} = \bar{N}^{FS}. \tag{31}$$

In conclusion, the production triangle equation represents a good start for analysing the interdependence between the most important KPIs. However, for more complex production models, the analytical verification of its validity is rather difficult and time consuming. That is why results based on simulation or data obtained from practice are more effective in obtaining the expended proof of this managerial relationship.

5 Improving Detailed Scheduling by Simulation

The internal orders released at the ERP level, due to their aggregated form, are not capable of being directly executed. This is especially true in the case of manufacturing orders (set of jobs) to be performed at workshop levels, where more detailed routings and schedules are needed. That is why the MES level aims at initiating detailed schedules that meet the master plan goals defined at the ERP level. The fine-scheduler receives the actual data of dependent orders, products, resource environment and other technological constraints (tools, operations, buffers, material-handling facilities and so on). The POM defines the manufacturing goals and their priorities. Obviously, the POM from time to time may declare various goals. The scheduler always has to provide a feasible sequence of jobs for machine lines or machining centres. As a result of the fine-scheduling process, a detailed production programme is obtained, which declares the releasing sequence of jobs and assigns all the necessary resources to them. It also proposes the starting time of the operations. At the same time, it must not break any of the hard constraints and has to meet the predefined goals.

The computation time of the detailed scheduling process is also an important issue to be taken into consideration, especially when a large number of internal orders, jobs, operations, resources, technological variants and constraints are involved.

Usually, all data of a specific manufacturing order (identifier, priority, product type, product quantity, due date, etc.) are available and can be downloaded from the ERP system database. On the other hand, all the related information on products, technology and resources (bills of materials, technological process plans, set-up times, processing intensities, NC programs, etc.) are available in the MES database.

The proposed fine scheduling approach combines model building, model simulation and model evaluation processes in two hierarchical layers, each layer having two phases (Fig. 10) [13]. Each layer presents a feedback loop aiming at improving the scheduling result and the model. However, for effectiveness and flexibility reasons, interactive human intervention has to be an option in both layers.

In the first phase (the top layer) the wide-range possible fine schedules space is scanned based on deterministic data models and on fast execution-based simulation to generate several, near-optimal feasible fine schedules. In the second phase (bottom layer) a more precise model-based event-driven simulation performs the sharp tuning.

In the top layer, the focus is set on creating only some near-optimal feasible schedules by considering detailed constraints and capabilities of resources, while in the bottom layer the model adjusting for various uncertainties and unpredicted events plays the primary role. In the bottom layer, the given set of a low number of potential fine schedules is evaluated according to viewpoints of stability and behaviour in an uncertain environment. That is, due to reduced number of possible trials, simulation is also performed rapidly, despite the fact that stochastic issues and unpredicted events can also be considered in the production model.

The proposed two-phase simulation-based model supports the flexible usage of multiple production goals and requirements simultaneously. The elaborated approach helps to solve the complex, detailed scheduling problem as a whole without its

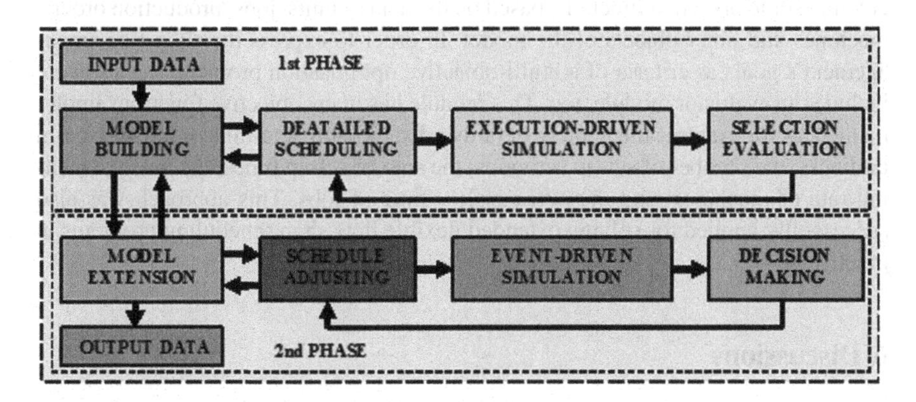

Fig. 10 Two-layered simulation-based fine scheduling

decomposition. In this way, all issues like batching, assigning, sequencing and timing are handled simultaneously in a much shorter time.

Moreover, simulation may easily lead to evaluation of several production performance indicators even before the manufacturing execution. That may serve for in-time corrections or compensations at the MES or even at the ERP level, with immediate results in both general and specific production performance.

Based on the above-mentioned two-phase modelling and simulation approach and in order to implement it in practice and solve the production fine scheduling problems, special developments had to be carried out. To implement the first phase (the top layer), a knowledge-intensive searching algorithm and the corresponding software were developed at the Information Engineering Department of the University of Miskolc. The algorithm is based on execution-driven fast simulation, overloaded relational operators and multiple neighbouring operators. The core of the scheduler engine implemented explores iteratively the feasible solution space and creates neighbour candidate solutions by modifying the actual resource allocations, job sequences and other decision variables according to the problem space characteristics. The objective functions concerning candidate schedules are evaluated by production simulation representing the real-world environment with capacity and technological constraints. Items, parts, units and jobs are passive elements in the execution-driven simulation, and they are processed, moved, and stored by active system resources such as machines, material handling devices, humans and buffers. The numerical tracking of product units serves the time data of the manufacturing steps. The simulation process extends the pre-defined schedule to a fine schedule by calculating and assigning the time data, too. Consequently, the simulation is able to transform the original searching space to a reduced space by solving the timing sub-problem. This part of the approach encapsulates the dependency of real-world scheduling problems. Successful adaptation of the approach into practice is highly influenced by the efficiency and speed of the simulation algorithm.

The performance analysis of the created fine schedule can be performed by calculating some objective functions based on the data of units, jobs, production orders, machines and other objects of the model. In order to express the shop floor management's goals as criteria of a multi-objective optimisation problem, the software includes an evaluator module, too. This module has many objective functions implemented, as for example the number of tardy jobs; the sum of tardiness; the maximum tardiness; the number of set-up activities; the sum of set-up times; the average waiting rate of machines; and the average flow time of jobs. This approach was also successfully applied for solving extended flexible flow shop scheduling problems in practice [14–17].

6 Discussions

Companies operating in discrete manufacturing appear and are modelled as complex, multi-functional production systems. Production planning and control becomes a difficult task for production systems even of low and medium size. The number of production entities (product items, jobs, machines/workplaces) and of human and technological resources, the logistical and technological constraints, the variations of operation routing and intensities, the variety of production goals to face today's economic and social challenges as well as the uncertainties of production processes and market demands make the control decisions very difficult.

Basically, control means decision-based acting and intervention to achieve the given goals. Control decisions are based on available information, which usually has to be acquired (directly and/or indirectly) and has also to be processed.

One indispensable requirement for control is its real-time character. That is, the control functions (information acquisition and processing, decision making and intervention) must be performed within given time intervals defined by the dynamic behaviour of the controlled processes. As general rule, the control actions must be faster than the rate of change of the controlled processes. In real-time control the sizes of the different time constraints may constitute a kind of hierarchical organising principle for the company control activities; the lower the control hierarchy is, the shorter the time interval available for performing the control functions.

At the execution level of technological processes real-time control is characterised by time limits of seconds or even fractions of a second. However, dedicated and embedded hardware and software systems can perform well due to the reduced number of state variables (physical quantities) they control.

The upper control level of MOM faces very strict time limits, too. The increased number of manufacturing orders and entities, of business goals, of logistical and distribution constraints, of operations routings and production goals, as well as the uncertainties of market demands and of failures of machining capacities make the everyday work of MOM experts as well of software developers very challenging.

At higher (ERP, MRP) hierarchical levels computer aided applications present more scalable models for helping functions of planning and control. At the MES level control decisions are more model-dependent, to such an extent that use of predefined or standard models is limited. This is probably the main reason why the up-to-date commercially available MES applications still need important customisation, and why the ISA-95 standard performs still below expected results. Practical experience shows that production control at the MES level performs proactively, by short-term detailed scheduling, and reactively, by on-line tracking, eventually re-planning tasks day by day. These activities can be carried out either by decisions of experts, or by using of integrated model-based decision support applications.

The qualitative assessment of the control activities can be evaluated from the quantitative results of production processes. In this respect, without continuous monitoring of some basic key performance indices (like throughput time, WIP level, utilisation of resources) the control of production is of lower quality. The dynamic equilibrium of the KPI values is influenced by appropriate control decisions to be taken before the execution phase of production processes.

7 Conclusions

Simulation models are suitable both for planning and evaluation of the performance of manufacturing systems. Based on practical experience, a two-phase hierarchical problem solving technique was proposed. In the first phase, at the top layer, creating jobs, allocating resources and filtering alternative feasible schedules is performed by a customised, very fast simulation model-based algorithm, which outputs only a few candidate schedules at the end. In the second phase, at the subordinated layer, these schedules are evaluated in more detail by an event-driven simulator that can include stochastic or uncertain elements. These supplementary features permit a deeper analysis of different variations on routing alternatives and production rates, for allocations of resources, as well as for priorities and goals. In this way, production management decisions support production results more effectively.

Acknowledgments The described work was carried out as part of the TÁMOP-4.2.2/B-10/1-2010-0008 project in the framework of the New Hungarian Development Plan. The realization of this project is supported by the European Union, co-financed by the European Social Fund.

References

1. Askin, R.G., Standridge, C.R.: Modeling and Analysis of Manufacturing Systems. Wiley, New York (1993)
2. Baker, K.R.: Introduction to Sequencing and Scheduling. Wiley Edition, New York (1974)
3. Bauer, A., Bowden, R., Browne, J., Duggan, J.: Shop Floor Control Systems—from Design to Implementation. Chapman & Hall, London (1993)
4. Bikfalvi, P., Erdélyi, F., Tóth, T.: A new approach to solving the tasks of production planning and control: the "Production Triangle" model. In: Proceedings of the XXV MicroCAD Conference, , pp. 7–14, Miskolc (2010)
5. Brucker, P.: Scheduling Algorithms. Springer, Berlin (2007)
6. Buzacott, J.A., Shanthikumar, J.G.: Stochastic Models of Manufacturing Systems. Prentice Hall, Englewood Cliffs (1993)
7. Erdélyi, F.: Modelling the performance of shop floor logistics by means of key performance indices. Paper presented at the 60th anniversary of the faculty of transportation of the Budapest University of Technology and Economics, Budapest, Hungary (2011)
8. Erdélyi, F., Tóth, T., Kulcsár, G.Y., Bikfalvi, P.: Some new considerations for applying MES models to improve the effectiveness of production operations management in discrete manufacturing. In: Proceedings of the 14th International Conference on Modern Information Technology in the Innovation Processes on the Industrial Enterprises—MITIP 2012, pp. 391–400, Budapest, Hungary (2012)
9. Gifford, C. (ed.): The Hitchhiker's guide to manufacturing operations management, instrumentation, systems, and automation society, ISA. http://www.isa.org/Template.cfm?Section=Books3&Template=/Ecommerce/ProductDisplay.cfm&ProductID=9217 (2007)
10. Instrumentation, Systems, and Automation Society: ISA95 manufacturing enterprise systems standards and user resources, 3rd edn. http://www.isa.org/template.cfm?section=standards2&template/ecommerce/ProductDisplay.cfm&ProductID=11366
11. Kiss, D., Tóth, T.: The methods of theoretical approach in production planning and control. In: Hetyei, J. (ed.) Information Systems for Enterprise Management in Hungary, pp. 59–94. Computerbooks, Budapest (in Hungarian) (1999)

12. Krajewski, J., Ritzman, B.: Operation Management (Strategy and Analysis). Addison-Wesley Publishing Co., New York (1996)
13. Kulcsár, GY., Bikfalvi, P., Erdélyi, F., Tóth, T.: A new simulation-based approach to production planning and control. In: Proceedings of the 37th International MATADOR Conference, pp. 251–254, Manchester, GB, 25– 27 July 2012
14. Kulcsár, G.Y., Erdélyi, F.: A new approach to solve multi-objective scheduling and rescheduling tasks. Int. J. Comput. Intell. Res. **3**(4), 343–351 (2007)
15. Kulcsár, G.Y., Erdélyi, F.: Modelling and Solving of the Extended Flexible Flow Shop Scheduling Problem, Production Systems and Information Engineering, vol. 3, pp. 121–139. University of Miskolc, Miskolc (2006)
16. Kulcsár, GY., Kulcsárné-Forrai, M.: Detailed Production Scheduling Based on Multi-Objective Search and Simulation, Production Systems and Information Engineering, vol. 6, pp. 41–56. University of Miskolc, Miskolc (2013)
17. Kulcsár, GY., Kulcsárné-Forrai, M.: Solving Multi-Objective Production Scheduling Problems Using a New Approach, Production Systems and Information Engineering, vol. 5, pp. 81–94. University of Miskolc, Miskolc (2009)
18. Little, J.D.C.: A proof of the Queuing formula $L = \lambda W$. Oper. Res. **9**, 383–387 (1961)
19. MES Centre of Excellence: 2012 MES product survey. http://www.mescc.com/participants.html (2013)
20. Monostori L., Váncza, J., Kis, T., Kádár, B., Viharos, Z.S.: Real-time cooperative enterprises. In: Proceedings of the 12th International Conference on Modern Information Technology in the Innovation Processes of the Industrial Enterprises—MITIP 2006, pp. 1–8, Budapest (2006)
21. Nyhuis, P., Von Cieminski, G., Fischer, A.: Applying simulation and analytical models for logistic performance prediction. CIRP Ann. **54**(1), 417–422 (2005)
22. Papadopoulos, H.T., Heavey, C., Browne, J.: Queuing Theory in Manufacturing Systems. Chapman & Hall, London (1993)
23. Pinedo, M.J.: Planning and Scheduling in Manufacturing and Services, 2nd edn. Springer, London (2009)
24. Quadt, D., Kuhn, H.: A taxonomy of flexible flow line scheduling procedures. Eur. J. Oper. Res. **178**, 686–698 (2007)
25. Scholten, B.: Integrating ISA-88 and ISA-95, Presented at ISA EXPO 2007, Houston, Texas. http://www.isa.org (2007)
26. Scholten, B.: MES Guide for Executives: Why and How to Select, Implement, and Maintain a Manufacturing Execution System, Instrumentation, Systems, and Automation Society, ISA. http://www.isa.org/Template.cfm?Section=Books3&Template=/Ecommerce/ProductDisplay.cfm&ProductID=10580 (2009)
27. Scholten, B.: The Road to Integration: A Guide to Applying the ISA-95 Standard in Manufacturing, Instrumentation, Systems, and Automation Society, ISA. http://www.isa.org/Template.Cfm?Section=Books3&Template=/Ecommerce/Productdisplay.Cfm&Productid=9212 (2007)
28. Tóth, T., Erdélyi, F.: New consideration of production performance management for discrete manufacturing systems. In: MITIP Conference, pp. 435–444 (2006)
29. Vernadat, F.B.: Enterprise Modelling and Integration. Chapman & Hall, London (1996)
30. Wang, W.: Flexible flow shop scheduling: optimum heuristics, and artificial intelligence solutions. Expert Syst. **22**(2), 78–85 (2005)

New Theory and Application for Generating Enveloping Surfaces Without Undercuts

László Dudás

Abstract The design and improvement of kinematical motion transfer surfaces, namely gear surfaces, require the modeling of the surface-surface enveloping process and the visualization of contact characteristics. To analyse the quality of mesh in respect to undercuts, this study uses the special visualisation capability of the Surface Constructor (SC) system, which is intended for creating enveloped contacting surfaces. This view uses the unique *Rho* = *Rho(Fi)* functions of the Reaching Model theory that have been applied earlier for different cases. Using these functions the evaluation and modification of mesh can be accomplished without the generated member of the surface pair. This study presents a new compressor type with a rotor that needs precise surface finishing. The patented grinding machine applied works with a special grinding wheel generated by the rotor surface. A novel undercut visualisation and detection method is used for the analysis of the correct grinding settings.

Nomenclature

e, f	$p1$, $p2$ parameter line indices
g, h	Indices of last $p1$, $p2$ parameter lines
is	Counter of intersected quadrangle edges
$k1, k2$	Contact lines
kx, ky	Constants
$p1, p2$	$F1$ surface parameters
$p1_e, p2_f$	$F1$ surface parameter values in an $F1$ grid-point, e =0, .., $g-1$; f =0, .., $h-1$
$(x0_{k,l}, y0_{k,l}, z0_{k,l})$	P_k point in the $K0$ frame, determined at (T_k, Z_l) grid point of $F2$

L. Dudás (✉)
Department of Information Engineering, University of Miskolc,
Miskolc-Egyetemváros, Miskolc 3515, Hungary
e-mail: iitdl@uni-miskolc.hu

G. Bognár and T. Tóth (eds.), *Applied Information Science, Engineering and Technology*, Topics in Intelligent Engineering and Informatics 7, DOI: 10.1007/978-3-319-01919-2_8,

$(x2_{k,l}, y2_{k,l}, z2_{k,l})$	P_k point in the $K2$ frame, determined at (T_k, Z_l) grid point of $F2$
A, B	Points on $F1$
$F1, F1_1, F1_2$	Generating surface, 1st and 2nd intersection
$F1(p1, p2)$	Generating surface given by surface parameters $p1$ and $p2$
$F2$	Generated surface
$F2_1, F2_2$	Surfaces generated by the same $F1$ surface
Fi, *Fi*1, *Fi*2	Φ variables of different surface generations
$Ki(xi, yi, zi)$	Coordinate system ($i = 0, 1, 2, 100$)
M, M_i	$F1$ surface points
M_t	$F1$ surface point corresponding to P_k
P_k	Contact point, $k = 0, .., r$
P'_k	Generated point of $F2$, $k = 0, .., r$
P_n	Point of singularity
R	Space coordinate of κ coordinate system and reaching direction, *mm*
$R(\Phi)$	Reaching-coordinate function at T_k division in κ_l coordinate system associated to Z_l, *mm*
R_u	R value of P_k point, *mm*
R_t	R value of M_t point, *mm*
$R_{k,l}$	Minimum R value determined at (T_k, Z_l) grid point of $F2$ *mm*
Rprev	R minimum determined in previous iteration cycle for same $T - Z$ grid-point, *mm*
$R(x0, z0)$	Changeover from the $(x0, z0)$ Descartes coordinate system to the curved $R - T$ coordinate system
ΔR	Difference between the results of the last two cycles
*Rho*1, *Rho*2	R variables of different surface generations
T, T_k	Division coordinate, $k = 0, .., r$, *mm* or *grad*
T_A, T_B	Division coordinate values at points A and B
Tau, *Tau*1, *Tau*2	T variables of different surface generations
$T(x0, z0)$	Changeover from the $(x0, z0)$ Descartes coordinate system to the curved $R - T$ coordinate system
$T1, T1_1, T1_2$	Body of the generating object, 1st and 2nd intersections
v, w	Degree of partial derivation
Z, Z_l	$F2$ grid-parameter value and identifying parameter of κ coordinate system at the same time, $l = 0, .., s$, *mm* or *grad*
Zeta, *Zeta*1, *Zeta*2	Z variables of different surface generations, *mm* or *grad*
$\boldsymbol{n}$	Surface normal vector
$\boldsymbol{r0}$	Vector form of of $F1$ in $K0$ coordinate system, $\boldsymbol{r0} = \boldsymbol{r0}(p1, p2)$
$\boldsymbol{r0}^{(h)}$	Vector form of $F1$ in $K0$ coordinate system given by symbolic algebraic expressions
$\boldsymbol{r1}^{(h)}$	Vector form of $F1$ in $K1$ coordinate system given by symbolic algebraic expressions
$\boldsymbol{r}_{AB}$	Array of crossing points of a $p1$ or $p2$ parameter line and the $F1$ intersection curve; given in $K0$ coordinate system
$\boldsymbol{v}^{1,2}$	Relative speed vector
$\boldsymbol{M}_{i,j}$	Homogenous transformation matrix from Kj to Ki coordinate system given automatically by symbolic algebraic expressions
$\kappa(\Phi, R, T)$	Non-Descartes coordinate system

κ_l	κ coordinate system, associated to Z_l $F2$ grid-parameter value
Φ, Φ_i	Motion parameter and space coordinate of κ coordinate system, $i = 0,..,p$, *mm* or *grad*
Φ_t	Φ value corresponding to P_k contact point, *mm* or *grad*

Remark: There is no difference between italic or normal form of above notations in the text.

1 Introduction

Most modern gear development software applies the Tooth Contact Analysis (TCA) method for optimizing the contacting properties of gears. This method involves the determination of the distance function—the ease-off topography—between the contacting tooth surfaces, and is equally suitable for conjugate and for modified, barreled or profile crowned gear surfaces. It can produce the transmission error and the bearing pattern, and is thus a very powerful tool (see e.g. [10, 11]). One of the best realizations uses real-time ease-off topography manipulation carrying out surface or kinematical system parameter alteration using a mouse and visual feedback [14]. However, such methods need the two surfaces to be previously computed, also in the case of conjugate contacting. The development in the space of motion tracks is a characteristic only of the Surface Constructor (SC) software application. Using this tool it is possible to optimize the contacting characteristics of the mesh and to avoid interference situations *without* the generated surface. Though many of the capabilities of SC can be found in other gear development tools (see the software comparison in Fig. 1), and especially in ZAKGEAR software, special features of SC include freedom in entering generating surfaces and kinematic relations due to the inbuilt symbolic algebraic computation capability and the software facilities based on the $R = R(\Phi)$ functions. As these functions are very important for the detection of undercut situations, they will be the focus in this chapter.

1.1 Short Introduction to the Theory of the Algorithm

To give a basis for referencing in the modeling section, we need to briefly review the theoretical fundamentals of the innovative Surface Constructor kinematical surface generating and contact analysis software. The name of the theory introduced by the author for generating conjugate surface pairs is the Reaching Model. The model solves the well-known task for gearings: the determination of the $F2$ conjugate surface if the generating $F1$ surface and the generating motion are given.

There are two well-known methods for this task:

- the differential-geometric method developed by *Gochman* [1], and

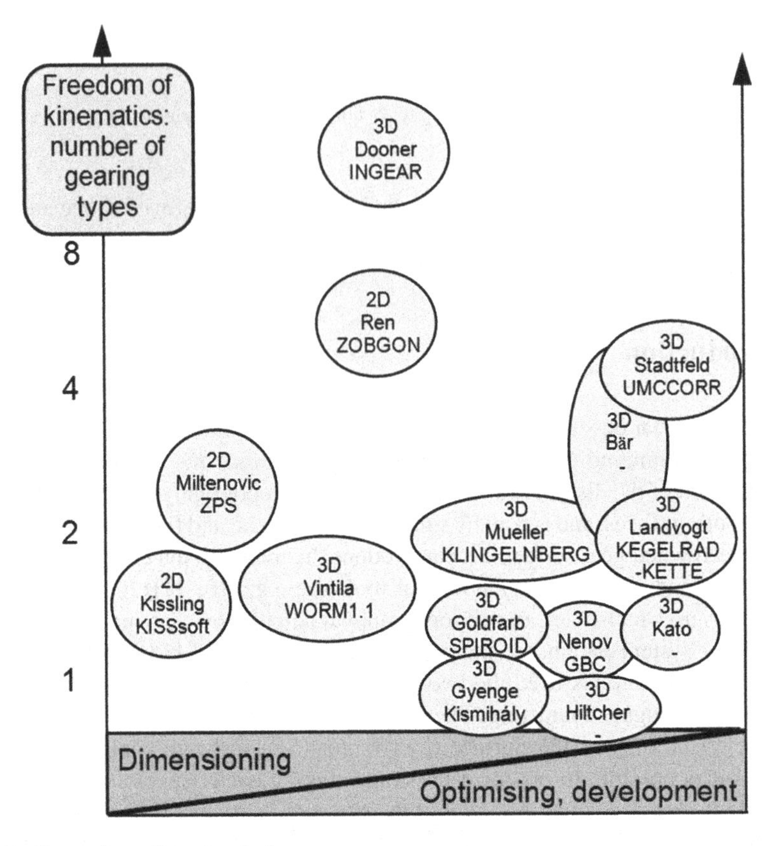

Fig. 1 Comparison of gearing design programs

- the kinematical method, which applies the $\boldsymbol{n} \cdot \boldsymbol{v}^{1,2} = 0$ scalar product where $\boldsymbol{n}$ is the normal vector and $\boldsymbol{v}^{1,2}$ is the relative velocity vector at the contacting point. This method has two forms: the pure geometrical type and the *Litvin* type, which uses matrix-algebra [9].

Before the Reaching Model, the word undercut was used for various undesirable surface problems of mating surface pairs. The Reaching Model includes the capability to detect all types of local undercuts and the global cut in the same theoretical model. For a detailed explanation see [5].

The main advantage of the Reaching Model is its simplicity. In this model the generation of one of the points of the $F2$ surface is equivalent to solving a simple minimum value problem. The denominative reaching process will be introduced briefly based on Fig. 2.

The Reaching Model applies a special non-Descartes coordinate system κ. The Φ coordinate lines as well as the coordinate axis Φ itself coincide with the motion paths of the points of a surface that is in the coordinate system $K2$, so Φ has two roles:

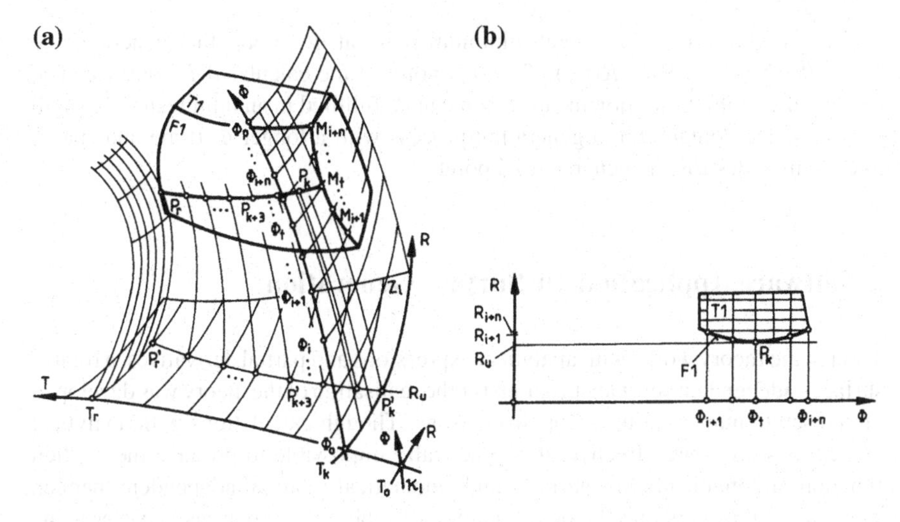

Fig. 2 The reaching process

1. motion (time) parameter,
2. one of the three space coordinates of coordinate system κ.

In the reaching process we choose a Φ coordinate line that does not intersect the $T1$ body. Stepping from one Φ coordinate line to another going in direction R, the generating Φ coordinate line will be that one to reach $F1$ first. This Φ line is the path of motion of point P_k' that will be one surface point of the generated surface $F2$. Point P_k will be the contact point.

The necessary condition of connection in the Reaching Model is

$$\frac{\partial R}{\partial \Phi} = 0 \tag{1}$$

where $R = R(\Phi, T, Z)$ is the reaching-coordinate function, and Φ is the space coordinate along the motion paths, T is the division coordinate in the κ slicing coordinate system, and Z is the identifying parameter of the κ coordinate system. This necessary condition is equivalent to the $\boldsymbol{n} \cdot \boldsymbol{v}^{1,2} = 0$ condition of the above-mentioned kinematical method.

The sufficient condition of the local minimum in the $\Phi = \Phi_t$ point, which is equivalent to the real mesh at the same location, given in the following form:

$$\frac{\partial^v R}{\partial \Phi^v}|_{\Phi = \Phi_t} = 0 \; (v = 1, 2, ..., w-1) \tag{2}$$

and

$$\frac{\partial^w R}{\partial \Phi^w}|_{\Phi = \Phi_t} > 0 \tag{3}$$

where w is an even number.

This condition defines a local minimum point at $\Phi = \Phi_t$ that generates the $P'_k = P'_k(\Phi = \Phi_t; R = R(\Phi_t); T = T_k)$ point of the calculated $F2$ surface. The local nature of this condition means that it can be fulfilled in an infinitesimally small time or space domain, but a generating process with a longer Φ (time and space) interval may destroy the generated P'_k point.

2 Software Application for Surface Generation

Though the theory works with analytical expressions and partial derivatives, a robust, surface-independent software program for the realisation of the theory was developed on a discrete numerical basis for two reasons. Though calculations with analytical expressions are more effective, it is practically impossible to produce the explicit functions automatically in a surface- and kinematical-relation independent manner. The second reason is that the determination of global cut situations is easier using discrete computer simulation of the enveloping process.

2.1 The Calculation Algorithm

The applied iterative algorithm to determine the P_k minimum point of the $F1$ intersection curve can be followed in Fig. 2. In this general example the applied discrete R coordinate lines drawn at $\Phi_i (i = 0, .., t, .., p)$ intersect the $F1$ surface at points $M_{i+1}, .., M_t, .., M_{i+n}$. The $M_t \equiv P_k$ point has the minimal R value among these points. The core of the algorithm determines these intersection points and the corresponding $R_{i+1}, .., R_t, .., R_{i+n}$ reaching coordinate values. Because of the discrete nature of the algorithm the exactness is not theoretical but the error can be controlled, as will be shown below. Calculating the R_i values for ascending Φ_i series makes it possible to detect not only minimum or maximum points, but also the locations of quasi-inflection places, thus detecting local undercut problems.

The algorithm developed, somewhat restricting the generality, applies a plane for slicing the space instead of a curved R-T coordinate surface. Using a plane carries some advantages, see [3].

Figure 3 shows a frequently used solution, which applies the simplest linear R reaching coordinate direction and linear T division coordinate direction in the $y0 = 0$ plane of a $K0$ Descartes coordinate system.

Such a $K0$ coordinate system has two roles in the Reaching Model and in the algorithm:

- It moves the $y0 = 0$ coordinate plane carrying the T and R directions, and creates the Φ motion tracks for every $(x0, z0)$—or$R$, T—point in the space, realising the fixed $\kappa = \kappa(\Phi, R, T)$ special curved coordinate system. The start position of the $y0 = 0$ coordinate plane is identical to the $\Phi = 0$ coordinate surface of the κ

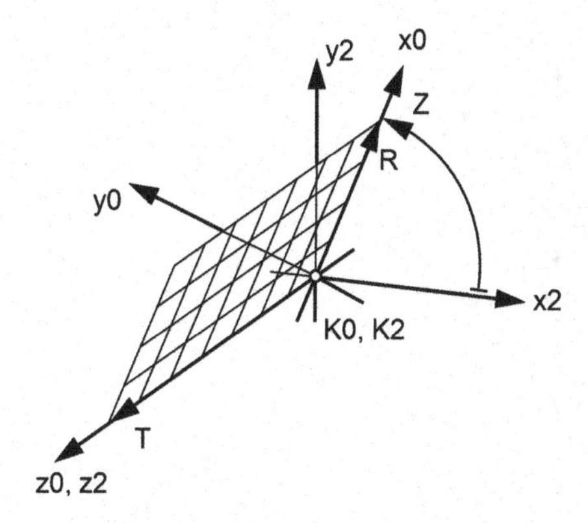

Fig. 3 Example giving the position of the slicing plane in the $K2$ frame with reaching direction R and T division coordinate in the plane $R = R(x0, z0) = x0$; $T = T(x0, z0) = z0$

coordinate system and will hold the determined $P_k', .., P_r'$ points as the points of the $F2$ required. Cf. Fig. 2.

- Because only one curve of the $F2$ searched surface appears in the $y0 = 0$ coordinate plane of the $K0$ coordinate system—and thus also in the $\Phi = 0$ surface of the κ system—the installation of further $K0$—and corresponding κ—coordinate systems is necessary. These coordinate systems are positioned by the Z parameter of the Reaching Model. The $y0 = 0$ coordinate planes of these coordinate systems, where $F2$ curves are searched for, have to slice the 3D space with single valued correspondence. To achieve this, the positive R half of the plane is used in the algorithm, as in Fig. 3.

The algorithm applies two transformation chains, one of which is between the fixed $F1$ and the moving $F2$ space: this is the $K1 - K2$ chain. The other gives the position of the $K0$ frame in the $K2$ as a $K2 - K0$ chain. Using these transformations — given by normal and inverse homogenous transformation matrices $\boldsymbol{M}_{1,2}$; $\boldsymbol{M}_{2,1}$ and $\boldsymbol{M}_{2,0}$; $\boldsymbol{M}_{0,2}$—the position of surface $F1$ can be determined in the $K0$ frame and the series of $\boldsymbol{M}_{i+1}, .., \boldsymbol{M}_t, .., \boldsymbol{M}_{i+n}$ points can be calculated. Using these points the grid points of surface $F2$ can be determined following the steps of the flowchart shown in Fig. 4.

After entering the given $F1$ surface, the relative motion of $F2$ space to $F1$ and the slicing motion symbolically, and then entering the numerical values, the Z cycle value l determines a slicing plane position. The T cycle value k selects an R reaching direction in the slicing plane. The inner Φ cycle determines the $(x0, z0) \equiv M_i$ intersection points and the corresponding R values and produces the $R_t \equiv R_{k,l}$ minimum value and the adjoined $M_t \equiv P_k$ point with its T_k and $R_{k,l}$ coordinates in the κ_l coordinate system. This core algorithm outputs $x0_{k,l}$, $y0_{k,l} = 0$ and $z0_{k,l}$ coordinates of P_k. Taking the coordinates of this point in $K2$ frame in the $\Phi = 0$ moment, the P_k' point is yielded. Repeating the T and Z cycles, the grid points of $F2$

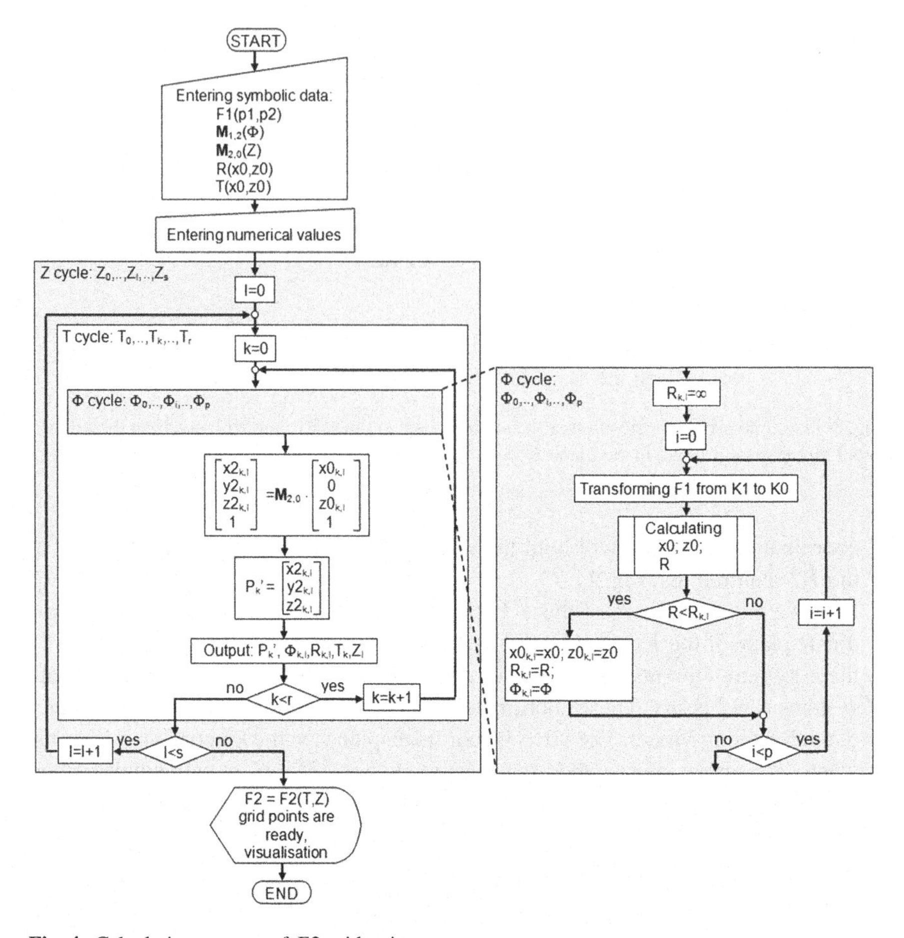

Fig. 4 Calculation process of $F2$ grid points

surface can be obtained. The points of a contact line are characterised by the same time, and thus by the same Φ value.

2.1.1 Determining the M_i Intersection Point

In the previous section it was shown that the determination of the M_i intersection point can be done in the $y0 = 0$ plane of $K0$ frame. If the coordinates of M_i point in $K0$ frame offer themselves as $x0 = x0_i$, $y0 = 0$, $z0 = z0_i$, then its coordinates in the κ_l coordinate system will be $\Phi = \Phi_i$, $R = R(x0_i, z0_i)$, $T = T(x0_i, z0_i)$. To realise this, the $F1$ surface that is given by $p1$ and $p2$ surface parameters in $K1$ coordinate system has to be transformed into the $K0$ coordinate system under fixed Φ_i and Z_l values.

If given the two-parametric vector form of $F1$ in $K1$ by homogenous coordinates:

$$\boldsymbol{r1}^{(h)} = \boldsymbol{r1}^{(h)}(p1, p2) = \begin{bmatrix} x1(p1, p2) \\ y1(p1, p2) \\ z1(p1, p2) \\ 1 \end{bmatrix} \tag{4}$$

then the vector form of $F1$ in $K0$ is:

$$\begin{aligned} \boldsymbol{r0}^{(h)} &= \boldsymbol{r0}^{(h)}(p1, p2, \Phi_i, Z_l) \\ &= \boldsymbol{M}_{0,2}(Z = Z_l) \cdot \boldsymbol{M}_{2,1}(\Phi = \Phi_i) \cdot \boldsymbol{r1}^{(h)} = \begin{bmatrix} x0(p1, p2, \Phi_i, Z_l) \\ y0(p1, p2, \Phi_i, Z_l) \\ z0(p1, p2, \Phi_i, Z_l) \\ 1 \end{bmatrix}. \end{aligned} \tag{5}$$

To follow the subsequent steps of the algorithm, look at Fig. 5. The figure represents not only the $p1$ and $p2$ parameter lines, but the reaching direction R at T_k coordinate value. For the M_i intersection point the R value is unknown in the κ_l coordinate system and so are the corresponding $x0$ and $z0$ coordinates in the $K0$ system. To determine these, the following successive approximation procedure is applied: first the algorithm seeks out that spatial quadrangle of the $p1$, $p2$ grid of $F1$ which is intersected by the R reaching direction. This quadrangle is determined by the $p1_e$, $p1_{e+1}$, $p2_f$, $p2_{f+1}$ parameter lines in the figure. The tiles are generally bordered by arches, but the approximation applies linear interpolation between grid points. The right quadrangle is determined by the situation when the intersection points of the quadrangle edges with $y0$ plane, A and B, embed the T_k value in the κ_l coordinate system by means of their T_A and T_B values.

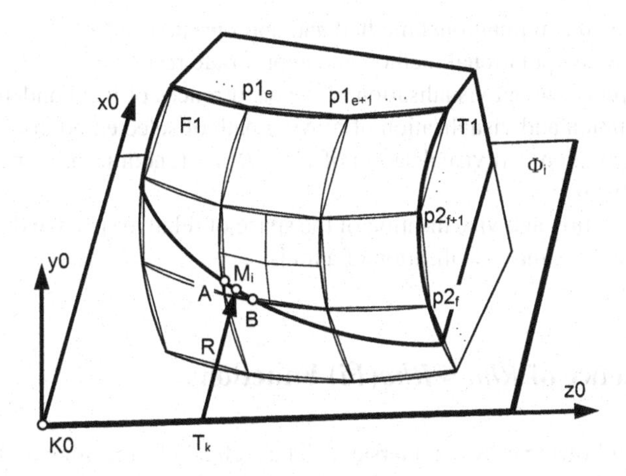

Fig. 5 The M_i intersection point determination

Applying linear interpolation between A and B points, the R value belonging to the T_k coordinate can be calculated from the expression used for defining the reaching direction. Because of linear approximation, this value may differ from the exact R value of the M_i point. To keep this difference within a given error interval, a more precise R value is calculated by refining the calculation using smaller quadrangles by halving the $p1$ and $p2$ parameter intervals. The approximation continues until the difference between the two last R values becomes smaller than the given error limit. The M_i point obtained in such a manner corresponds to the required P_k contact point if its R coordinate is the smallest among those offered by the different discrete Φ_i values of the Φ interval. The flowchart of this approximation is shown in Fig. 6. By increasing the number of Φ steps in the Φ interval, the error caused by the discrete enveloping simulation can be decreased.

2.2 The Surface Constructor Application

The system developed applies both symbolic and numerical representations of the objects. The symbolic algebraic computation provides the flexibility characteristic of the tool. Surface Constructor starts as an empty kinematical modelling shell and models the kinematical modelling process itself (see [6]). The system sketched in the lower right corner of Fig. 6 has three main representation levels:

- the symbolic level, which uses a symbolic algebraic representation of the objects in the kinematical model,
- the numerical level, which stores the given and computed objects using numerical form, and
- the visualisation level, which allows the analysis of views and motion of the objects.

The selection and visualisation options are as follows:

- F2glob: global computational method and appropriate result
- F2lok: local computational method and appropriate result
- F2al: computation and visualisation of the occurrences of local undercuts
- Φ: computation and visualisation of moving path of selected points
- R-Φ: computation and visualisation of $R = R(\Phi)$ functions as a special feature of this software
- v_a: computation and visualisation of the space of relative speed and acceleration
- PT: computation and visualisation of axoids.

3 The Essence of *Rho* = *Rho(Fi)* Functions

To make the following easy to understand, let us refresh the meaning of the function. The *Rho* is the reaching direction, a free plane curve, but in these examples lines will be applied for the sake of simplicity. It can be interpreted as a very thin telescope in

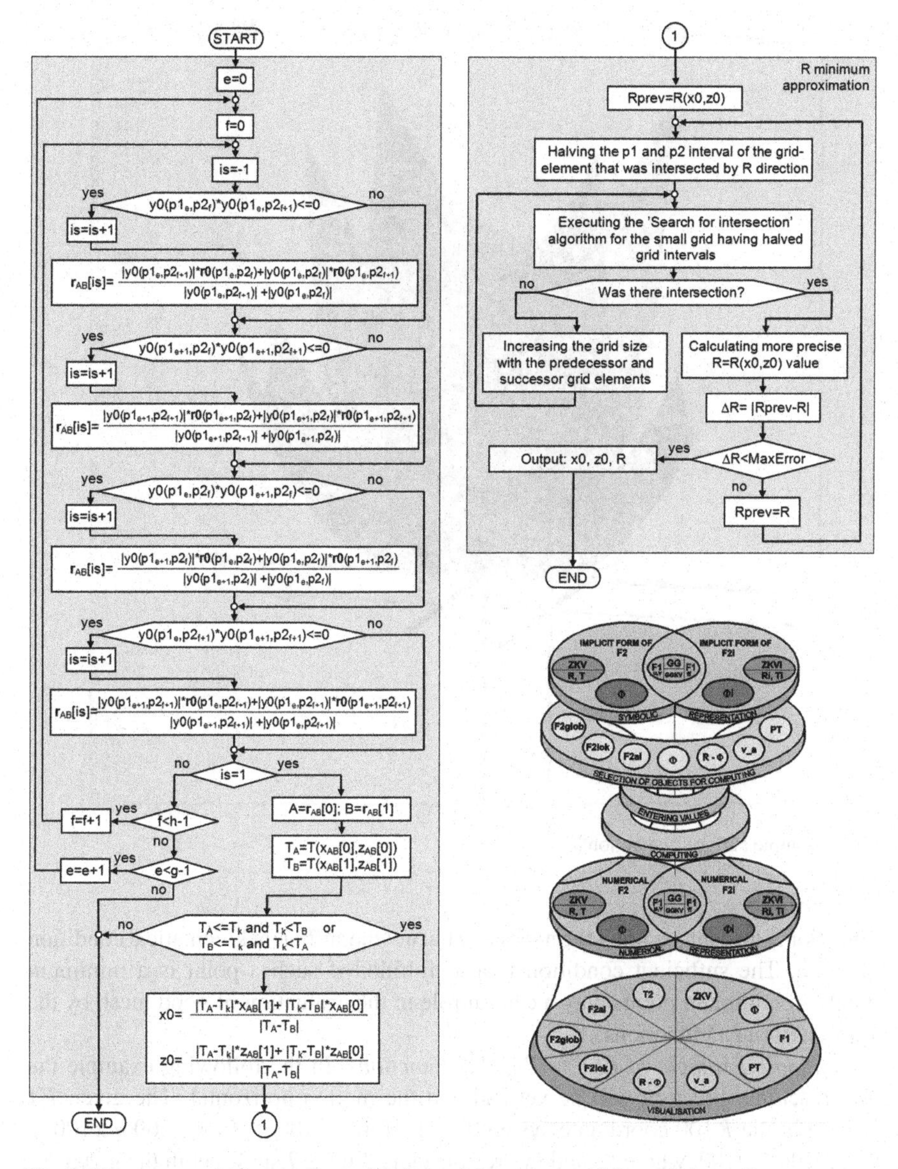

Fig. 6 The process of M_i point determination and the structure of the realising software

the space of the required second surface $F2$. As the given surface $F1$ moves in this space, it compresses and releases the telescope, so the length of the telescope changes in time. This $length = length(time)$ function corresponds to the $Rho = Rho(Fi)$ function. Instead of time, usually the motion parameter Fi is used. The minimum value of the length plays an important role because this creates a surface point of $F2$. Points of the function having a horizontal tangent—minimum, maximum,

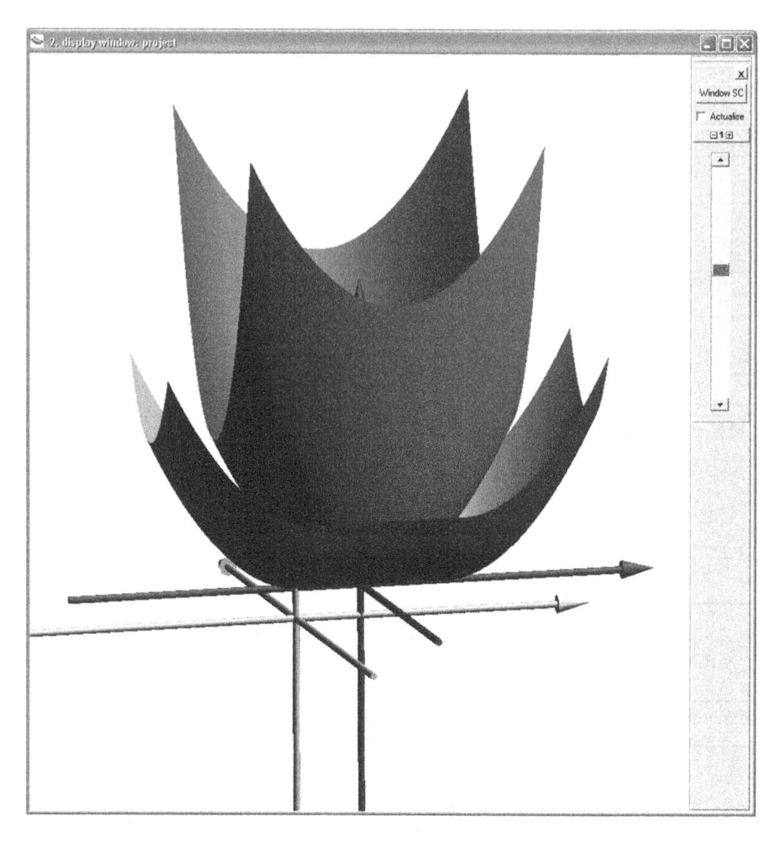

Fig. 7 Simple surface generation

inflection, constant—satisfy the necessary but not the sufficient kinematical condition of mesh. The sufficient condition is also fulfilled if such a point is a minimum point. So minimum points play a central role in the evaluation of good mesh by the $Rho = Rho(Fi)$ functions.

A Simple Example of Rho = Rho(Fi) Functions. In the following example the *Rho* reaching direction will be vertical and the motion horizontal. The given $F1$ surface in the $K100$ coordinate system has the form: $z100 = kx * x100 * x100 + ky * y100 * y100$, where *kx* and *ky* are constants. Figure 7 shows both the given $F1$ and the generated $F2$ surfaces. The boat-form $F2$ has a middle enveloped part and two footprint-like parts at the ends. Such footprint-like areas are not advantageous in gears mesh. In the Surface Constructor system there is a $Rho = Rho(Fi)$ function visualisation window that maps the generated $F2 = F2(Zeta, Tau)$ parametric surface into its own rectangle area. Moving the sliders or simply the mouse cursor in the window, the (*Zeta, Tau*) parameter regions can be scanned and for every (*Zeta, Tau*) pair, as well as for every $F2$ surface point, a $Rho = Rho(Fi)$ function can be shown. The *Zeta* direction is parallel to *Fi* direction, *Tau* is perpendicular

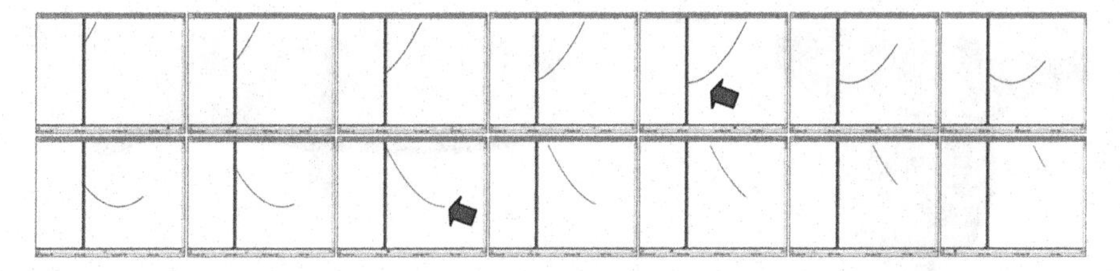

Fig. 8 $Rho = Rho(Fi)$ functions at $Tauindex = 100$

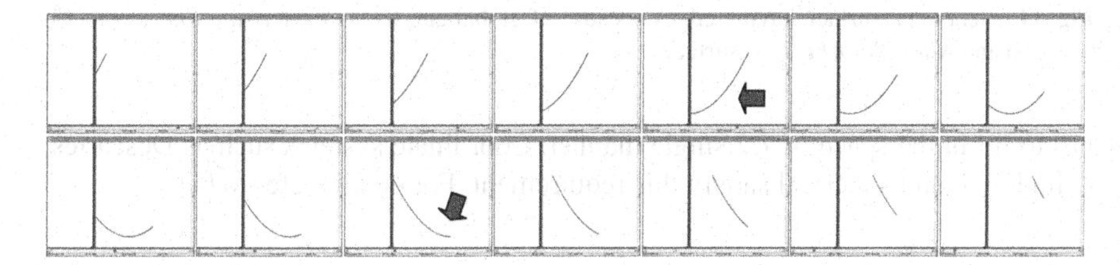

Fig. 9 $Rho = Rho(Fi)$ functions at $Tauindex = 50$

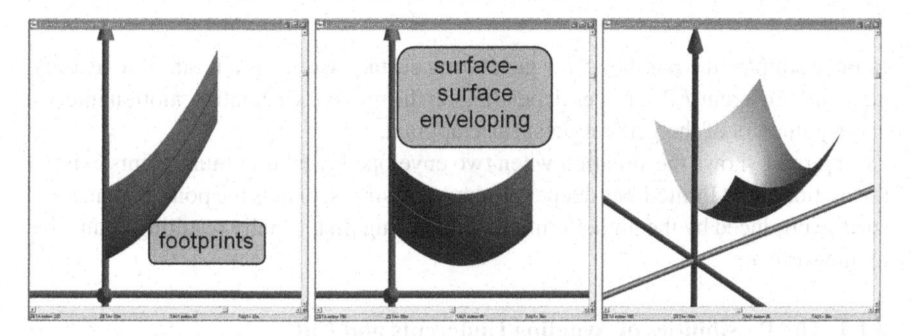

Fig. 10 $Rho = Rho(Fi, Tau)$ surfaces at $Zeta = 20$ and $Zeta = -50$

to the *Zeta* and *Rho* directions in this example. Figure 8 gives window shots for $Tauindex = 100$ and Fig. 9 for $Tauindex = 50$. Functions not having minimum with the horizontal tangent indicate footprint-like contact. Arrows emphasize window shots where surface-surface enveloping starts and ends ($Zeta = 0$; $Zeta = -100$).

It is also possible to plot the *Rho* = *Rho*(*Fi*,*Tau*), *Zeta*=*const* and the *Rho*= *Rho*(*Fi*, *Zeta*), *Tau* = *const* surfaces, as shown in Fig. 10. The valley form indicates surface-surface enveloping.

It is important that the characteristics—horizontal tangent, minimum point, inflection, horizontal line segment—indicating different contacting situations that are the same both for simple and complex relative surface motions do not depend on the complexity of shape of the given generating surface. To realise this advantage of the *Rho* = *Rho*(*Fi*) functions, the set of carrying planes of *Rho* reaching directions

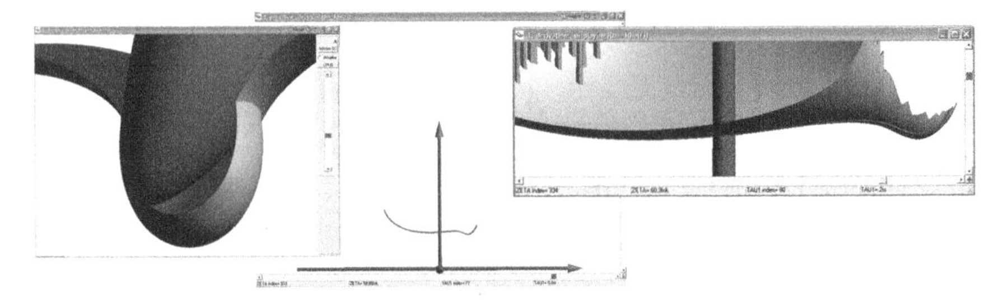

Fig. 11 Generated surface having edges because of global cut, the characterising *Rho* = *Rho*(*Fi*) function and *Rho* = *Rho*(*Fi, Tau*) surface

has to fill in the space of *F*2 single manner. Coordinate systems such as Descartes, cylindrical, and spherical satisfy this requirement. For details refer to [4].

3.1 Examples on Detection and Elimination of Undercuts

In the examples the parabolic *F*1 generating surface works as a tooth of a gear and generates different *F*2 surfaces depending on the relative kinematics, motion interval and parameters of the generating surface applied.

Figure 11 shows the situation when two enveloped local minimum points exist. If the motion is not limited, the deeper minimum destroys, or cuts the point of generated surface produced by the higher minimum occurring first, usually resulting in an edge on the surface.

3.1.1 The Possibilities of Avoiding Undercuts and Cut

The undercut is an infinitesimal happening opposite to the cut which requires a given space and time interval to occur. Local undercuts are usually followed and hided by global cuts.

The First Possibility: Limitation of the Generated Surface, so the Zeta Interval. This is a restrictive method and its use is limited, because the limited surface area does not have the same functionality and loadability as the original surface area. Figure 12 shows an example.

The Second Possibility: Limitation of the Motion, so the Fi Interval. A better method is to limit the motion interval in order to prevent an undercut region from appearing. This modification may result in a larger loadable, usable area of the generated surface because by limiting the undercutting motion the maximum of the generated enveloped surface can be saved. In the course of demonstrating this, let us stop for a minute to introduce the detection of the inflection type local undercut. This type of undercut occurs at a point that is more difficult to detect. Figure 13 shows the situation when the generating motion is limited so the dangerous global cut cannot fully appear. (The *Zeta* interval was also limited by half.) Scanning the *Zeta-Tau*

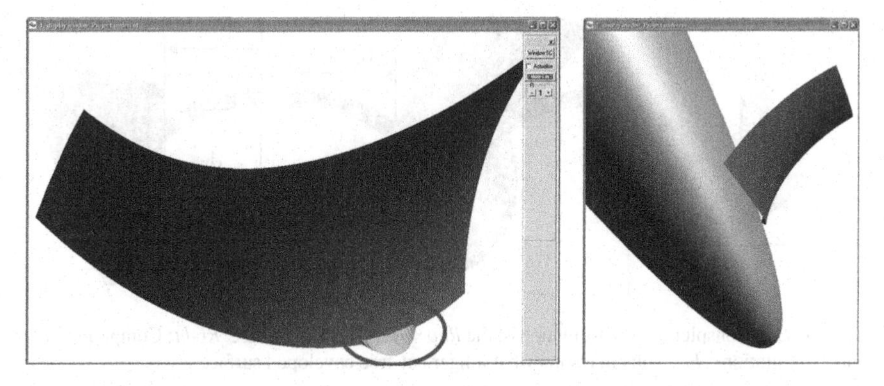

Fig. 12 *Left* There is a small remained undercut part after limiting *Zeta* = −90... +90 to *Zeta* = +60... +90. *Right* There is not undercut at *Zeta* = +61... +90

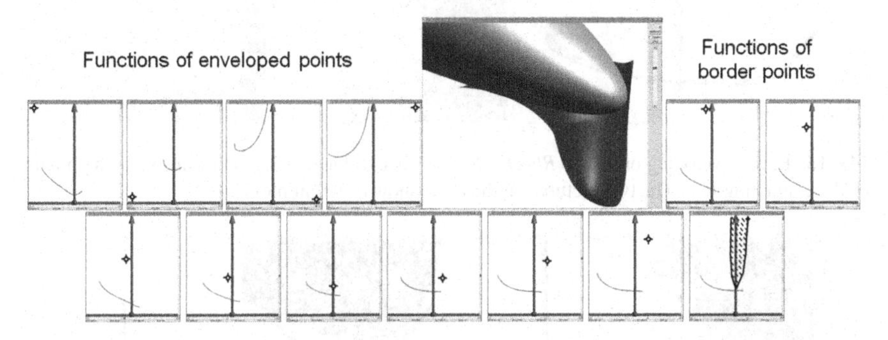

Fig. 13 Limited motion interval. Detection of border points of not enveloped region (hatched)

parameter region, the appearing inflections can be caught. Following these points, the border of *Zeta-Tau* region of enveloped points can be drawn. The outer points characterised by clear enveloping while the inner ones produced by cut or footprint formation.

To emphasise the advantage of *Fi* limitation opposite to *Zeta*, a generating surface in a deeper position was applied and caused a large undercut. Comparing the effects of the *Zeta* and *Fi* limitations it can be proved that *Fi* limitation provides a larger enveloped surface, see Fig. 14.

The Third Possibility: Modification of Kinematic Relation of Motion. By positioning the $F1$ surface to a higher position in the relative rolling motion, the resulting undercut becomes smaller. At a given parameter setting, enveloping of an edge can be observed: the sliding velocity of contacting point on $F1$ becomes zero for a short time, indicated by horizontal *Rho* = *Rho*(*Fi*) function segment, see Fig. 15.

The Fourth Possibility: Modification of the Generating Surface F1. Applying smaller *kx* and *ky* constants at $F1$ it becomes more flat. There is a little edge enveloping at *kx* = 0.02; *ky* = 0.008. Decreasing the values, the edge can be eliminated as shown in Fig. 16.

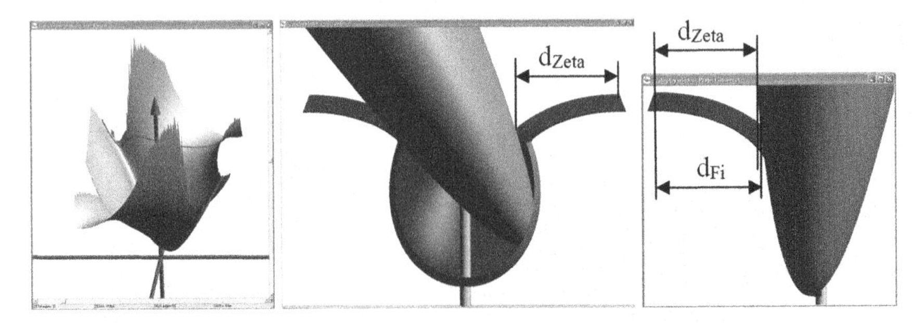

Fig. 14 *Left* Conspicuous cut formation of the *Rho* = *Rho*(*Fi*, *Tau*) surface. *Right*: Comparing *Zeta* and *Fi* limitations: *Fi* results in a somewhat longer cut-free enveloped surface

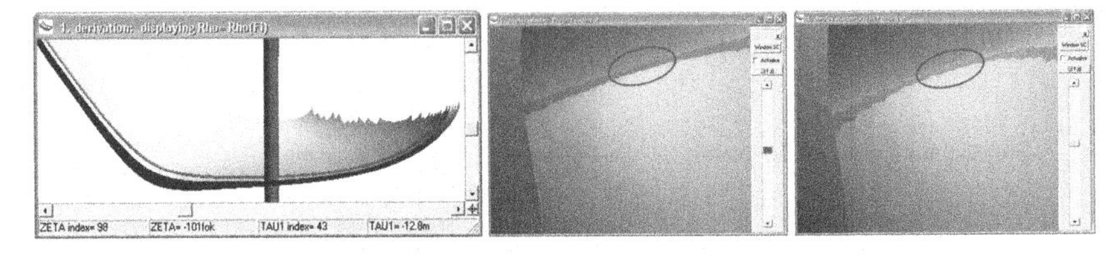

Fig. 15 Edge enveloping on *Rho* = *Rho*(*Fi*, *Tau*) surface. Edge enveloping means stopping points of the contacting *line*. The two pictures on the *right* show different moments

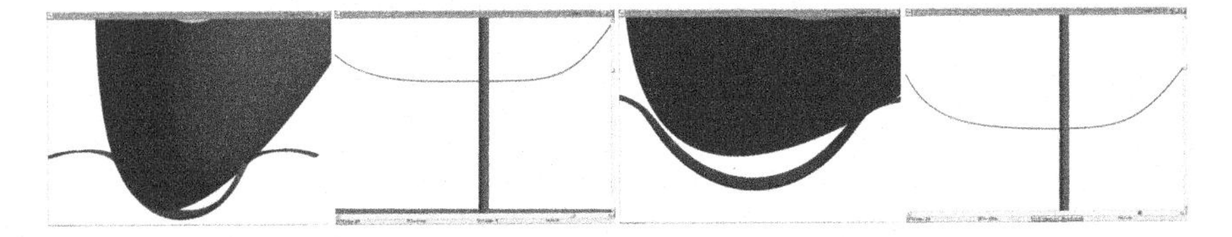

Fig. 16 Edge enveloping at $kx = 0.02$; $ky = 0.008$. Using smaller values at $F1$ results in $F2$ without an edge. The horizontal segment of *Rho* = *Rho*(*Fi*) function disappeared

4 Grinding Wheel Surface Generation

In this section a demonstrative example often use of the Surface Constructor for determination of a special grinding wheel surface enveloped by the rotor surface of a new type of compressor will be given. The grinding of this non-regular helical surface requires a non-conventional grinding machine and technology [8] to which the author holds a patent [2].

The result of the pump innovation is a new construction that has rotary parts only. The design method can be extended to compressors and expansion machines if the spiral angle of the helical surface is not equable. These new constructions, patented in recent years [13], are characterized by a rotary housing and rotor that form closed spaces or cavities. As the rotary parts rotate these cavities move parallel to the axes of the rotating machine elements. The required gap between the housing and the rotor has to be as small as possible to provide good sealing. This is essential for pumps and

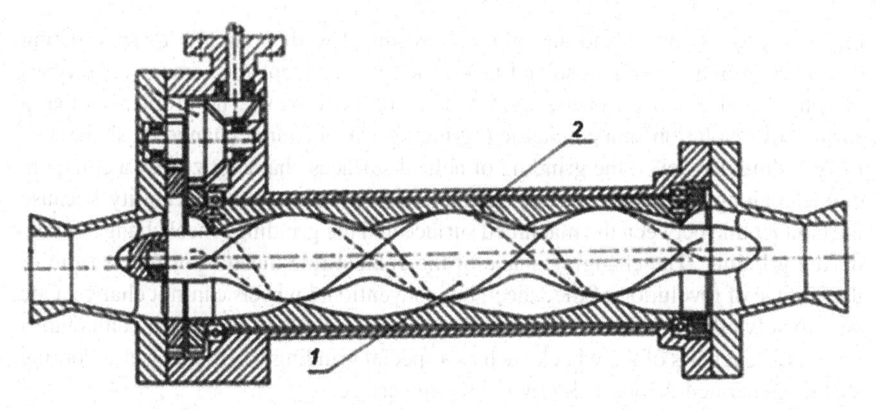

Fig. 17 FORCYL type water pump [12]

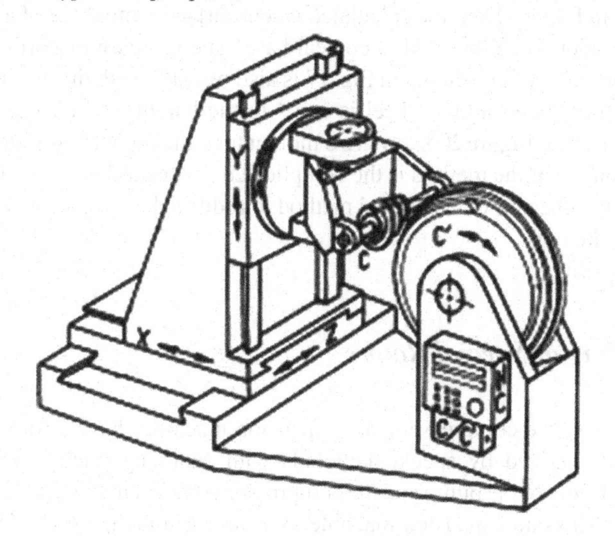

Fig. 18 New grinding machine construction for grinding non-conventional helical surfaces [2]

for expansion machines that work with steam energy. The airtight sealing is achieved by thick-film lubrication and the dynamic inertia effect of the oil [15]. The best way to ensure an airtight seal is to grind the working surfaces, but this raises problems. In the case of fluid pumps with normal helical surfaces, as shown in Fig. 17, the solution for rotor *1* is well known and is similar to the grinding of worms. The grinding of rotary house *2* is more complicated because of the inner surface, but is also possible [3]. The determination of grinding wheel profile is possible using modern software, like HeliCAD [3] or Surface Constructor [6, 7], and the profile of the determined surface of a revolution grinding wheel can be dressed by CNC dressers.

The grinding of helical surfaces having a non-constant lead requires the special technology and machine introduced in [3]. This machine has a constant wheel/workpiece ratio. The grinding machine shown in Fig. 18 is also needed for the finishing process of spiroid and globoid worms. The theoretically exact grinding of such worms was an unresolved task in the past, except for some special types of spiroid and globoid worms, e.g. involute spiroid worms. The problem is that a surface of revolution shaped classical grinding wheel cannot change its shape during grinding. However, the grinding of helical surfaces characterized by a changing diameter along the threads or a having non-constant lead requires this ability, because the contact line between the machined surface and the grinding wheel changes shape during grinding. The changing shape of the contact line means a different form of the surface of revolution of the wheel, but conventional wheels can-not change. One way to solve this problem is if the different surface parts of the worm can connect with different parts of the wheel, such as a special grinding wheel that has a working surface generated theoretically by the worm surface.

The modeling of the grinding of a spiroid worm with the new machine construction is presented in Fig. 19. Here the calculated wheel surface reminds us of a Reishauer-type gear-grinder wheel but with a conical base. The grinding of normal helicoids like the rotor of the pump shown in Fig. 17 is also possible with the new technology. As the width of the grinding wheel is equal to the length of the rotor there is no need for axial feed. Figure 20 shows two moments of the modeled grinding process. The disadvantage of the method is the complicated production process of the wheel. Because of this, the non-conventional method introduced here can be used efficiently for mass production.

4.1 Grinding of Special Rotors

The new, innovative compressors and expansion machines have a rotor and rotary housing characterized by special helicoids with changing pitch. The difference between the rotors of the pump and the compressor is shown in Fig. 21. The compressor can be used as an expansion machine by changing the direction of the rotation of the elements. As the construction of such machines is fairly new, first a short introduction of the construction is given. The rotor and the rotary housing have their own, parallel axes.

Figure 22 shows the rotor and the opened rotary housing. Between the two axes there is a distance. The two parts rotate around their individual axes in the same direction and form cavities in every moment. These cavities move parallel to the axes, along the length of the machine. In the case of a compressor or an expansion machine these cavities become smaller or larger, respectively, as the parts rotate. The compression or expansion ratio depends on the pitch variation along the axis, not on the angle velocity of the rotation. The grinding of the rotary housing is problematic because of the length and relatively small diameter. Possibilities and limitations require further analysis, not presented here.

The grinding of the rotor is also problematic because of the undercut possibility. From earlier experiments on the grinding of worms having constant pitch it seems to be evident that the grinding wheel tilt angle has to be set close to the lead of the

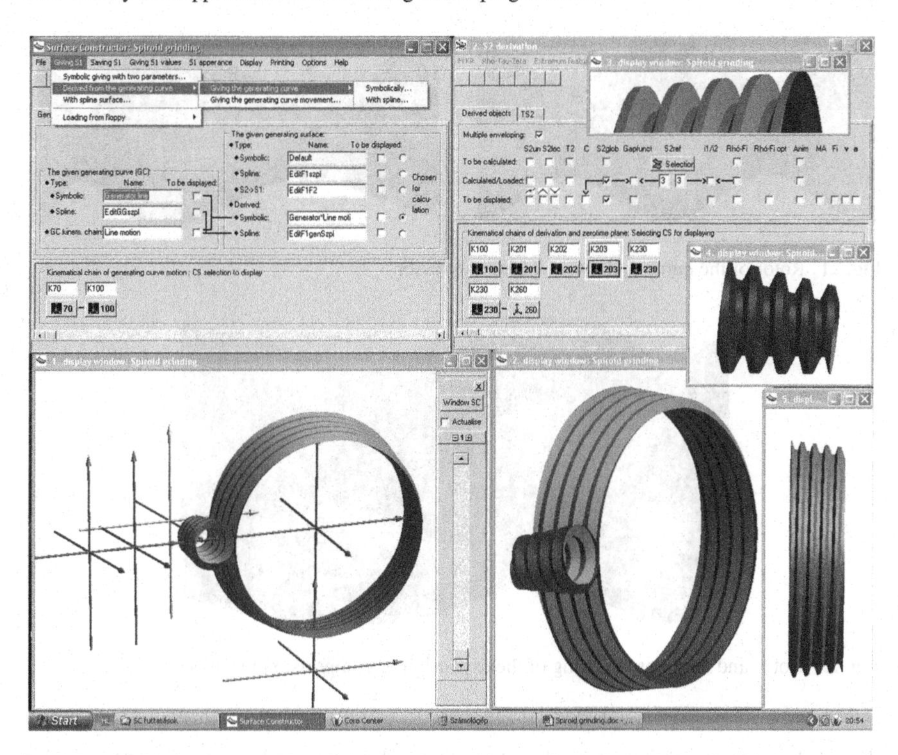

Fig. 19 Generation of grinding wheel for spiroid worm

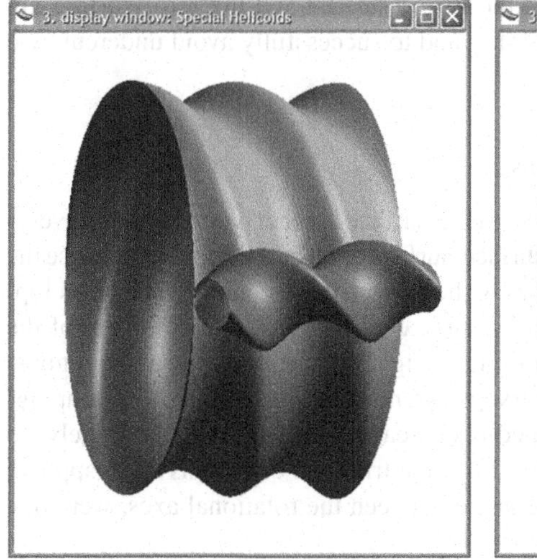

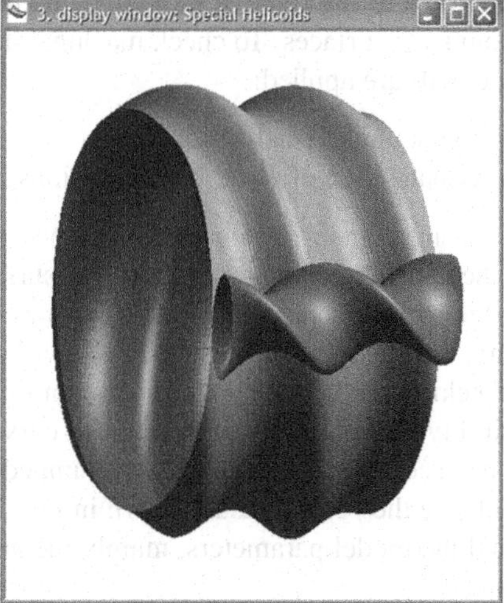

Fig. 20 Full-width grinding of the rotor

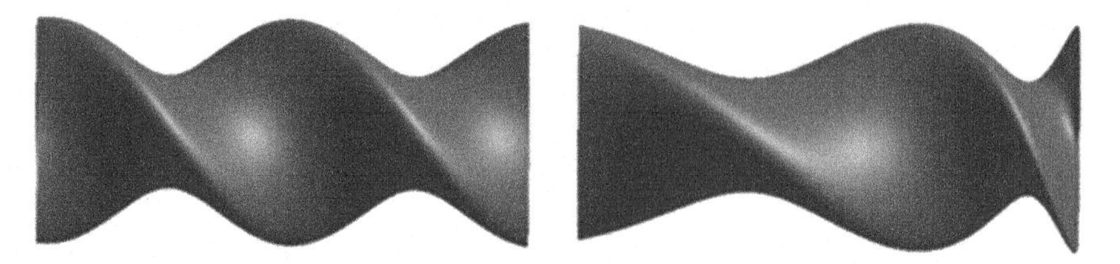

Fig. 21 Rotor of the pump and rotor of the compressor

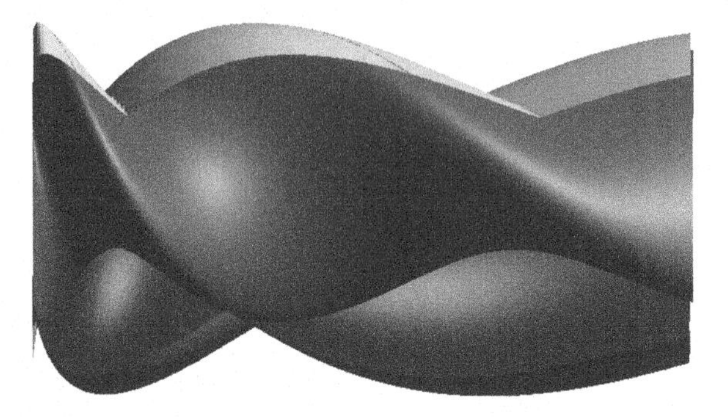

Fig. 22 Rotor and the rotary housing of the expansion or compression machine

thread that has the smallest pitch value. To model the grinding process the special gear CAD software named Surface Constructor (SC) is used.

SC generates the grinding wheel surface as an enveloped surface, generated by the rotor in the relative motion. Thus the wheel surface and the rotor surface are conjugate surfaces. To check machinability and to successfully avoid undercut, two methods are applied:

- contact line visualization and
- visualization of $R = R(\Phi)$ functions.

Using contact line visualization, the surface-surface contact has to be found everywhere in the enveloping process. The surface-surface contact is important because the edge-surface contact may mean generating the surface by sweeping, not by enveloping, or an edge can be enveloped by the surface of the rotor. One moment of the checking can be seen on the right-hand side of Fig. 23. Because of the undercut possibility the continuity of the contact curve is essential. A discontinuity or 'jumping' on the contact line means that the jumped-over area will not be ground accurately. To achieve the perfect result shown in Fig. 23, some trial-and-error runs were applied, and the model parameters, mainly the angle between the rotational axes, were fine tuned.

The problem with contact line visualization is that it has to be applied in every moment of the rotation. As Fig. 23 shows, the checking requires proper angle and zoom settings at every moment. Though the user interface of the visualization is very

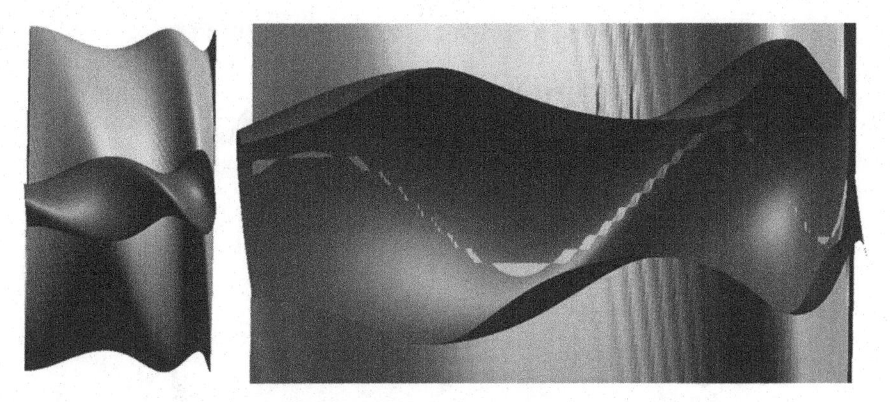

Fig. 23 Calculated grinding wheel surface (*left*) and checking of perfect grinding by the contact line (view from the inside of the rotor) (*right*). Gamma = 5°

user friendly, the checking process is time consuming. Fortunately SC has a unique visualization capability.

The exclusive capability of the Surface Constructor software, as well as the visualization of the $R = R(\Phi)$ functions, were discussed in the previous section. The visualization of the $R = R(\Phi)$ functions uses the relative kinematical relation of the rotor and the wheel as basic information and optimizes the grinding wheel surface and its setting in the space of relative velocity. Instead of relative velocity vectors it uses the motion tracks because these lines are tangential to the velocity vectors.

4.2 Undercut Control by the R = R(Φ) Functions

In most cases the given generating surface is changeable, or the kinematical relations have some flexibility. In these situations we can optimize the contacting properties by changing the value of one of the parameters of the generating surface or the kinematical system. In the particular situation when the derivation of the grinding wheel surface is required without undercut problems, the main parameter that can be varied is the angle between the rotation axes. The distance of these axes, and thus the grinding wheel diameter, can also be modified, but angle modification is preferred.

The first idea was that the best angle is the smallest lead angle of the rotor, which appears where the pitch is the smallest. This means an angle between the axes of about Gamma = 5°. The calculations showed good contact and enveloping surface by surface using the contact line visualization method for checking against undercut. This can be seen in Fig. 23. However, using the second checking method the visualization of the $R = R(\Phi)$ functions gave a more precise solution. The more accurate investigation showed that there is an undercut problem at the $T = 0$ coordinate indicated by the inflexion form, as can be seen in Fig. 24. The T axis coincides with the wheel axis and Z angle is measured like in Fig. 3. It was somewhat

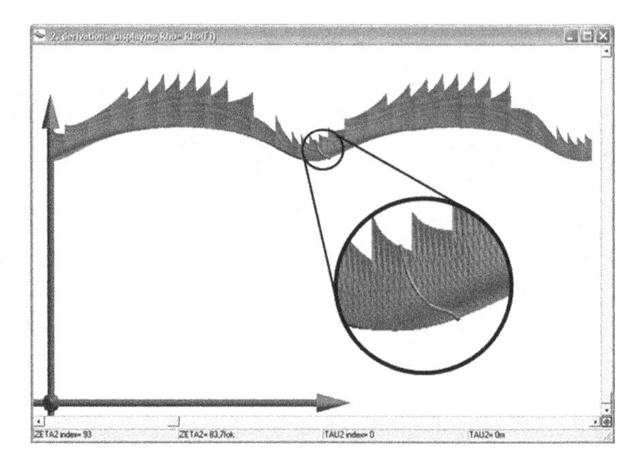

Fig. 24 $R = R(\Phi)$ function at Gamma = 5° indicates a problem in grinding wheel generation

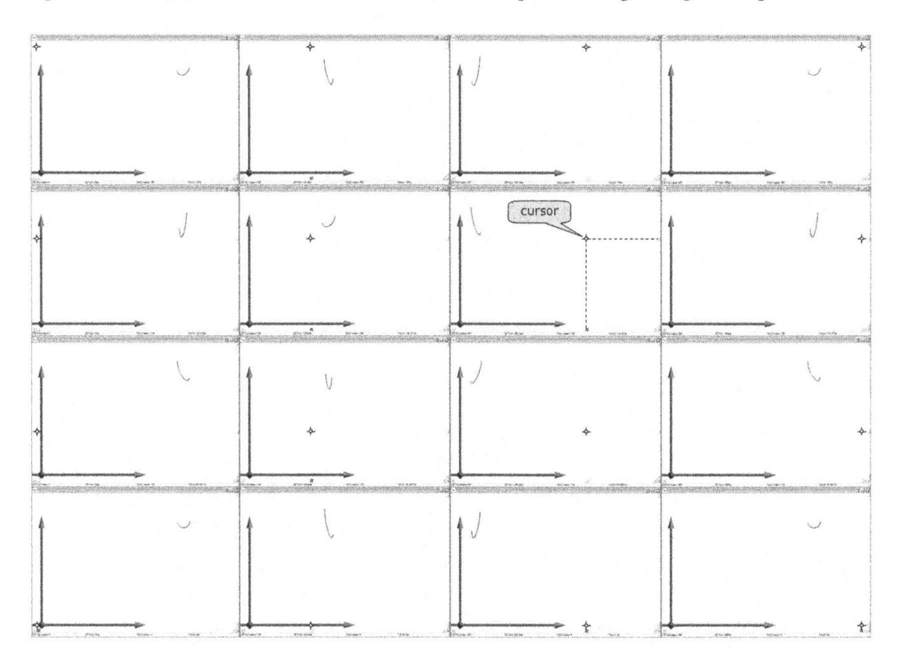

Fig. 25 $R = R(\Phi)$ functions at Gamma = 0°

surprising that the best result appeared at Gamma = 0°, as proven by Figs. 25 and 26. Figure 25 shows the $R = R(\Phi)$ functions in some peripheral and internal locations of the investigation area. The cursor position determines the T and Z values. Figure 26 shows all function curves as a surface and proves that every curve has a smooth minimum point, so the curves form a valley shape. This means that all the generated points of the grinding wheel surface are produced by surface-surface enveloping, so there is no undercut.

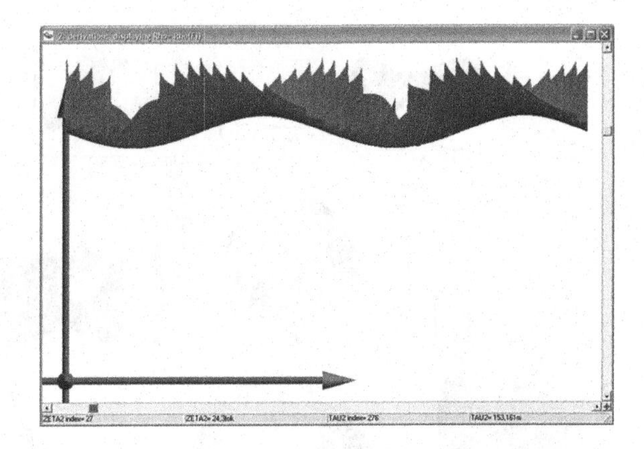

Fig. 26 Surface of $R = R(\Phi)$ functions at Gamma = 0°

The investigations proved that the given rotor can be grinded accurately. Unfortunately, the production of such a grinding wheel is expensive, and its use is reasonable in the mass production of rotors only. The suitable type of grinding wheel is not a dressable sort, but another, coated by an abrasive layer.

The performance of the investigations was aided by the Surface Constructor kinematical modeling and surface generating software tool. In the investigation process the computer program modeled the rotor surface, then the first generating process calculated the surface of the rotary housing, after which the second generating process determined the special grinding wheel. The generating processes applied enveloping of the calculated surface by the rotor surface. During the investigation the visualization of the surface objects helped check the results and monitor the modeling process. The software handles a maximum of eleven visualization windows at the same time. The content of any window can be freely animated, zoomed, or rotated. This freedom and the ease of entering the generator curves, e.g. ellipse for the rotor, and entering the kinematical relations using symbolic expressions made the investigation process convenient.

One screenshot of the investigation screen is given in Fig. 27. The upper part shows the generating surface entering window with the rotor, the first derivation window with the derived rotary housing and the second derivation window with the generated grinding wheel surface. The lower part of the screen shows additional visualization windows.

5 Summary

This chapter presented a new theory for the determination of enveloped surfaces. An algorithm that realizes the theory was introduced in detail. A short description was given of the Surface Constructor application that applies the algorithm and is suitable for the determination of enveloped surfaces and for detecting undercut problems.

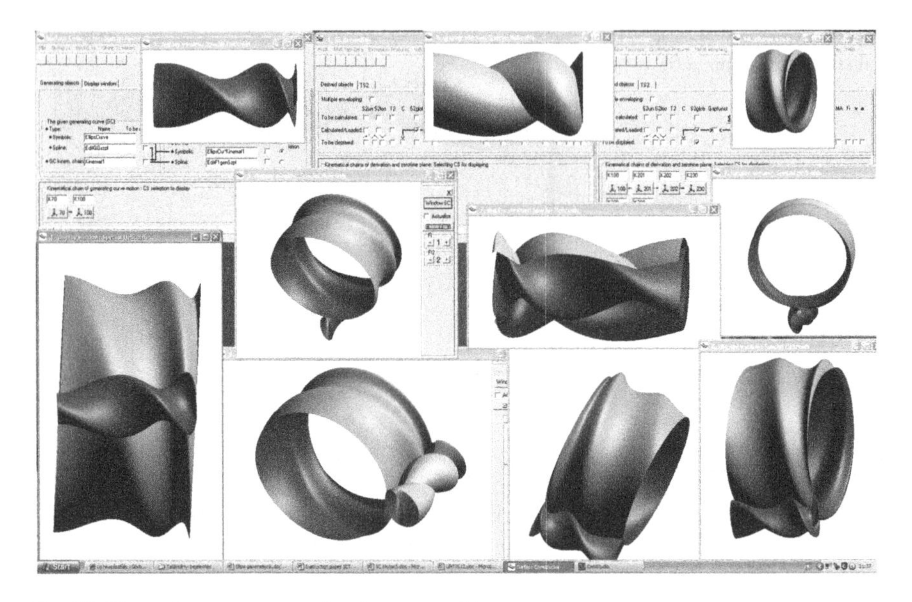

Fig. 27 The surface constructor development tool in use

The unique $R = R(\Phi)$ function visualization capability makes it possible to reveal the problematic generated surface regions and helps to avoid undercut problems. The study analyzed the grindability of the rotor of special pumps, innovative compressors and expansion machines from a geometrical point of view. The conclusion is that pump machine elements can be ground without a problem, and that the special rotors with a changing pitch for compressors or for expansion machines can be ground using a special machine, grinding wheel and technology. However, grinding of the inner surface of a rotary housing raises problems and requires further investigation and analysis. The investigation was performed by the Surface Constructor kinematical simulation and surface modeling tool, which has proved its applicability many times. The discrete simulation of the generating process offers an advantage when the determination of the real surface is required, but has a disadvantage when mathematical analysis is needed.

Acknowledgments The described work was carried out as part of the TÁMOP-4.2.2/B-10/1-2010-0008 project in the framework of the New Hungarian Development Plan. The realization of this project is supported by the European Union, co-financed by the European Social Fund.

References

1. Dudás, L.: Theory and Practice of Worm Gear Drives. Penton Press, London (2000)
2. Dudás, L.: Grinding machine, for grinding non-surface of revolution surfaces, especially conical and globoid Worms (in Hungarian). Hungarian patent HU P9003803 (1992)

3. Dudás, L.: Kapcsolódó felületek gyártásgeometriai problémájának megoldása az Elérés Model segítségével, Resolution of geometrical problems of contacting surfaces using the reaching model (in Hungarian). PhD Thesis, Hungarian Academy of Sciences, Budapest (1992)
4. Dudás, L.: Surface constructor—a tool for investigation of gear surface connection. In: Skolud B., Krenczyk D. (eds.) Proceedings of CIM: Advanced Design and Management Conference. Wydawnictwa Naukowo-Technicne, Warsaw, pp. 140–147 (2003)
5. Dudás, L.: Modelling and simulation of a novel worm gear drive having Point-like Contact. In: Proceedings of TMCE: Symposium. Ancona, Italy, pp. 685–698 (2010)
6. Dudás, L.: Advanced software tool for modelling and simulation of new gearings. Int. J. Des. Eng. **3**, 289–310 (2010)
7. Dudás, L.: Gear Investigations based on surface constructor kinematical modelling and simulation software. In: Proceedings of UMTIK: 14th International Conference on Machine Design and Production. & Prod., Güzelyurt, T.R. Northern Cyprus, pp. 731–742 (2010)
8. Dudás, L.: Determining grinding tools for special helical workpieces. In: Proceedings of the 13th International Conference on Tools ICT: Miskolc. Hungary, pp. 97–102 (2012)
9. Litvin, F.L.: Gear Geometry and Applied Theory. Englewood Cliffs, Prentice Hall (1994)
10. Lunin, S.: Interactive visualization with parallel computing for gear modeling. http://www.zakgear.com/Parallel.html, Accessed 19 June 2012
11. Lunin, S.: New Discoveries in WN gear geometry. http://www.zakgear.com/WN.html, Accessed 19 June 2012
12. Micro Europe Kft.: A Sokszögmegmunkálás Élvonalában (In the frontline of polygon surface machining) (In Hungarian). http://www.microeurope.hu/indexsziv.html, Accessed 19 Feb 2012
13. Murrow, K.D., Giffin, R.G.: Axial flow positive displacement turbine. U.S. Patent. 2009/0226336 A1 (2009)
14. Stadtfeld, H.J.: The universal motion concept for Bevel Gear Production. In: Proceedings of 4th World Congress on Gearing and Power Transmissions. CNIT-PARIS, France, pp. 595–606 (1999)
15. Stosic, N., Smith, I.K., Kovacevic, A.: Opportunities for innovation with screw compressors. Proc. IMechE, J. Proc. Mech. Eng. http://www.staff.city.ac.uk/ ra601/oportsvi.pdf. Accessed 19 Feb 2012

New Solutions in Online Sheet Thickness Measurements in Incremental Sheet Forming

Imre Paniti

Abstract This work discusses an approved analytical framework of Single Point Incremental Forming (SPIF) of sheet metals, which is capable of modelling the state of stress in the small localised deformation zone in case of corners, flat and rotationally symmetric surfaces. The discussion focuses on the investigation of the sheet thickness prediction in the shell element used in the framework. Novel solutions are introduced in terms of on-line sheet thickness measurement and an adaptive control algorithm in SPIF. Results were experimentally verified for 0.5 mm thick Al 1050 sheets with variable wall angle geometry. Furthermore, a new device for two sided incremental sheet forming is introduced as a recently patented solution in Incremental Sheet Forming (EP2505279).

Nomenclature

F	Reaction force
F_x	x-component of the reaction force
F_y	y-component of the reaction force
F_z	z-component of the reaction force
r	Radial coordinate
r_{tool}	Radius of the forming tool
r_1	Radius of curvature of meridian at the shell element (equal to r_{tool})
r_2	Radius of the element normal where it cuts the z-axis
T	Measurement time
t	Thickness of the sheet
t_0	Initial thickness of the sheet
U	Hall voltage

I. Paniti (✉)
Institute for Computer Science and Control, Hungarian Academy of Sciences,
Budapest Kende u. 13–17, 1111, Hungary
e-mail: imre.paniti@sztaki.mta.hu

G. Bognár and T. Tóth (eds.), *Applied Information Science, Engineering and Technology*,
Topics in Intelligent Engineering and Informatics 7, DOI: 10.1007/978-3-319-01919-2_9,

Greek Symbols

α Defining the angle in the meridional plane at the shell element
β Wall angle
ϵ_θ Circumferential strain
ϵ_φ Meridional strain
μ_θ Circumferential component of the coefficient of friction
μ_φ Meridional component of the coefficient of friction
σ_m Mean stress
σ_t Thickness stress
σ_Y Yield stress
σ_θ Circumferential stress
σ_φ Meridional stress

1 Introduction

The study and the improvement of Incremental Sheet Forming (ISF) have been of considerable interest during the last couple of decades. The interest in these scientific investigations is because this process is flexible (the geometry of the formed part can be changed by the control of the machine without changing the tool) and enables higher formability as compared to conventional sheet forming technologies like stamping or deep-drawing. ISF is considered to be used only for small series and Rapid Prototypes, but the applicability of this process has a wide range in the fields of the automotive industry, aircraft industry, architectural engineering and medical aids manufacturing.

The basic idea of ISF is similar to metal spinning and goes back to a patent granted in 1967 by Leszak [12]. As opposed to metal spinning, where the sheet metal is formed into an axial symmetric object by being turned on a lathe and being pressed continually with a tool, in ISF the sheet is clamped down with a frame and formed with a flat or spherical head forming tool. In contrast to conventional plastic deformation processes the final shape of the part is not determined by specific dies, rather by the three-dimensional movement of a forming tool, for which an appropriate tool path has to be established, based on the part geometry. This movement can be carried out by a CNC milling machine or by an industrial robot. ISF can be divided into two main groups [10] depending on the number of contact points between sheet, forming tool and die (see Fig. 1).

In Single Point Incremental Forming (SPIF) the first contour of the part is supported by a backing plate without any specific die. In case of Two Point Incremental Forming (TPIF) a full or partial die is applied with a downward movement of the clamping frame [17].

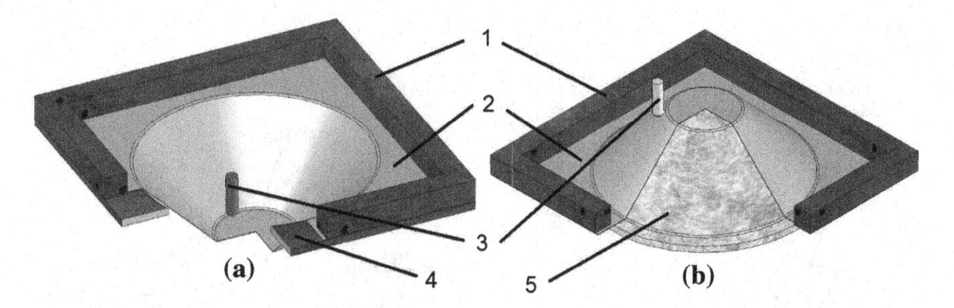

Fig. 1 **a** SPIF and **b** TPIF where *1* clamping frame, *2* sheet, *3* forming tool, *4* backing plate, *5* full or partial die

2 Analytical Background of SPIF

The analytical model introduced by Silva et al. [20] is based on membrane analysis with bi-directional in-plane contact friction forces to study the influence of major process parameters and their interaction.

This model uses the following simplifications:

- thickness, meridional and circumferential stresses assumed to be principal stresses;
- bending moments are neglected;
- axial symmetry is assumed;
- the material is simplified to be isotropic and rigid-perfectly plastic;
- the resultant friction stress which acts at the tool-sheet contact interface is assumed to consist of a meridional and a circumferential in-plane component.

Silva et al. [20] examined the smear-marks intervention between the forming tool and the surface of the sheet and concluded that the instantaneous small plastic zone of rotational symmetric SPIF can be approximated by a local shell element CDEF, illustrated in Fig. 2.

Figure 3 shows the acting stresses of the shell element in a schematic section view cut by an axial meridional plane. Figure 3a shows the top; and b the detail view. The

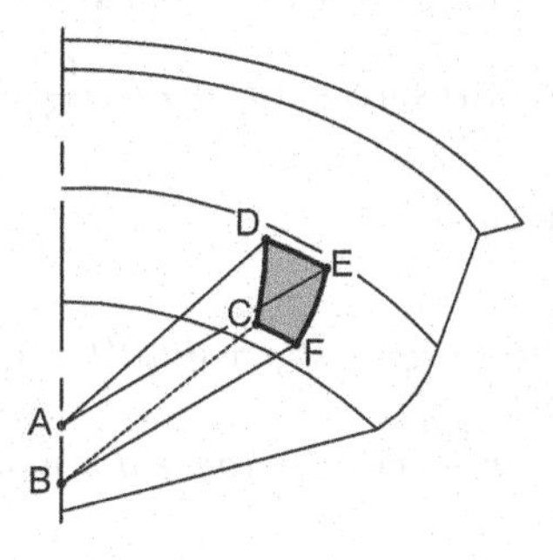

Fig. 2 Schematic perspective drawing of the shell element [20]

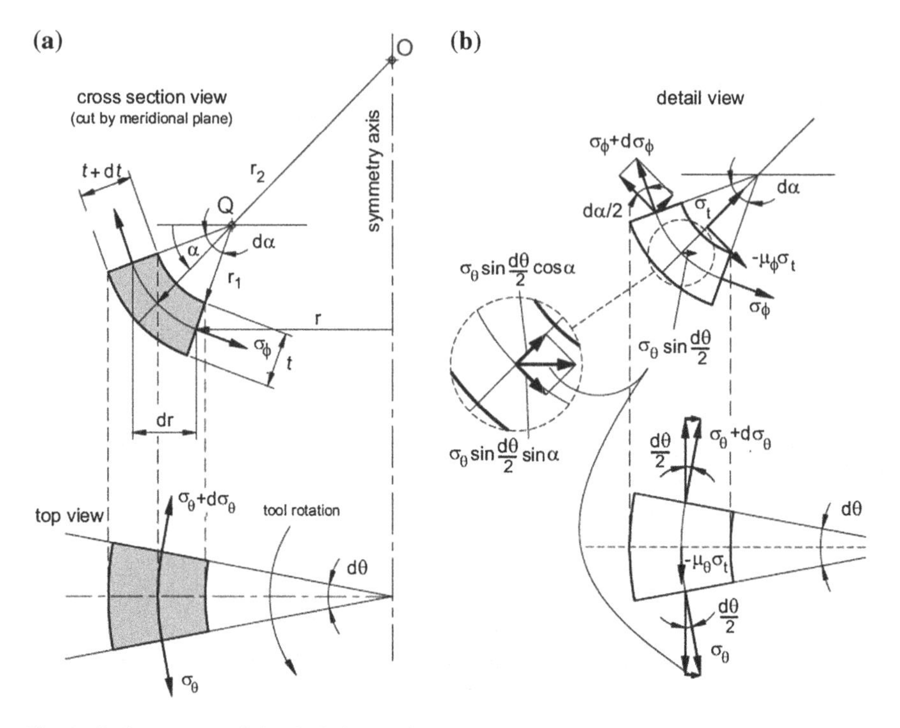

Fig. 3 Acting stresses of the shell element in a schematic section view cut by an axial meridional plane, with **a** *top view*, and **b** *detail view* [20]

normal of the local shell element cuts the z-axis at the point O and r_2 is defined as the corresponding radius. Point Q represents the centre of curvature on the normal and r_1 is the radius of curvature of the local shell element along the meridian direction [20].

By resolving the force equilibrium along the meridional, thickness and circumferential direction the following equations can be written (see [20]).

Along the meridional direction:

$$\begin{aligned}(\sigma_\varphi + d\sigma_\varphi)(r + dr)d\theta(t + dt) - \sigma_\varphi r d\theta t + \mu_\varphi \sigma_t r d\theta r_1 d\alpha \\ - \sigma_\theta \frac{d\theta}{2} r_1 d\alpha t \sin\alpha - (\sigma_\theta + d\sigma_\theta)\frac{d\theta}{2} r_1 d\alpha t \sin\alpha = 0\end{aligned} \tag{1}$$

Along the thickness direction:

$$\begin{aligned}\sigma_t r d\theta r_1 d\alpha + \sigma_\varphi r d\theta t \sin\frac{d\alpha}{2} + (\sigma_\varphi + d\sigma_\varphi)(r + dr)d\theta(t + dt)\sin\frac{d\alpha}{2} \\ + \sigma_\theta r_1 d\alpha t \sin\frac{d\theta}{2}\cos\alpha + (\sigma_\theta + d\sigma_\theta) r_1 d\alpha t \sin\frac{d\theta}{2}\cos\alpha = 0\end{aligned} \tag{2}$$

Along the circumferential direction:

$$\sigma_\theta r_1 d\alpha(t + \frac{dt}{2}) - \mu_\theta \sigma_t r_1 d\alpha(r + \frac{dr}{2})d\theta - (\sigma_\theta + d\sigma_\theta) r_1 d\alpha(t + \frac{dt}{2}) = 0 \quad (3)$$

Furthermore the model assumes that the flat and rotationally symmetric surfaces are formed under plane-strain conditions, $d\epsilon_\theta = 0$, and corners are formed under equal biaxial stretching, $d\epsilon_\varphi = d\epsilon_\theta > 0$. Under these conditions, considering the Tresca yield criterion—which states that $\sigma_1 - \sigma_3$ should be constant and equal to the flow stress of the sheet—the distribution of stresses in the small localised plastic zones can be obtained from Eqs. (1), (2) and (3) after neglecting higher order terms and taking into account geometrical simplifications [21].

In the case of flat and rotationally symmetric surfaces the mean stress can be written in the following simplified form:

$$\sigma_m = \frac{\sigma_Y}{2}\left(\frac{r_{tool} - t}{r_{tool} + t}\right) \quad (4)$$

For corners the mean stress can be written in the following simplified form:

$$\sigma_m = \frac{2\sigma_Y}{3}\left(\frac{r_{tool} - t}{r_{tool} + 2t}\right) \quad (5)$$

3 Sheet Thinning in SPIF

In 2009, Silva et al. [21] compared the analytical model described in the previous section with FEM simulations and experiments, and results showed good qualitative agreement. However, the thickness of the sheet is approximately estimated by means of the sine-law, originated from the spinning process [5];

$$t = t_0 \sin(90° - \beta). \quad (6)$$

The parameter t_0 is the initial thickness of the sheet and β is the wall angle (defining the angle between the un-deformed and the deformed sheet). In 2004, Young and Jeswiet [25] analysed radial profiles of truncated cones with various wall angles formed by SPIF. Results showed that sheet thinning in SPIF does not always follow the sine-law; the formula can only be used in case of low strains and in some cases the estimation of (6) is higher than in reality.

In 2010, Bambach [6] summarised different versions of the sine-law, e.g. where the thickness change is calculated as a function of the area ratio of the un-deformed and deformed triangular or quadrilateral mesh elements, and proposed a geometrical model of the kinematics of ISF for the prediction of membrane strains and sheet thickness. This model gives more accurate results than the sine-law, as it corrects the sine-law in curved areas and reproduces it on flat surfaces. However, the model has some limitations too.

Bambach emphasised that "due to the fact that material behaviour and friction are not included in the model it will not give accurate results in cases in which the sheet thickness is affected by the material behaviour in reality, e.g. for large wall angles where necking can occur or in situations dominated by friction" [6, p. 1571].

4 Sheet Thickness Measurements in SPIF

Due to the above mentioned deficiencies in the prediction of the sheet thickness in SPIF it is important to apply an on-line measurement method as a supplement. Based on the on-line thickness measurement an adaptive control could be applied to delay or even avoid fracture.

Some stereo vision systems are capable of measuring on-line the surface strains, so thickness strains can be estimated, but experimental investigations showed that only initial stages can be analysed with one paint layer [9]. The use of a paint layer with a pattern is necessary in order to trace the movements of material points under SPIF.

The experimental investigations of Van Bael et al. [24] and Aerens et al. [2] showed that the steady-state value of the axial reaction force during the SPIF process is proportional to the wall angle and to the resulting wall thickness. These force measurements in ISF are usually carried out using a three-component piezoelectric dynamometer, but other solutions like the application of piezo-resistive or strain gauged sensor system is also suitable. It is worth mentioning that forming force estimation can be carried out also from servomotor torque measurements taking into account the mechanical equilibrium of the machine tool structure used for the ISF process [18].

In 2006, Ambrogio et al. [3] claimed that the punch force trend is suitable as a "spy variable" of approaching material failure in the case of excessive sheet thinning. However, the previously mentioned measurement solutions can only provide indirect approximations as opposed to direct sheet thickness measurements.

In 2010, Dejardin et al. [7] also emphasised that the on-line monitoring of sheet thinning is important for the industrialisation of SPIF and for the process optimisation. They introduced a new solution for in-process thickness measurements based on the integration of an ultrasonic thickness measurement device into a forming tool. Dejardin et al. reported a measurement accuracy mean equal to $\pm$ 0.01 mm with proper repeatability [7]. However, one disadvantage of this approach is that the immersion transducer requires a propagation media with a specific viscosity. An accurate measurement could only be obtained with a 70 % water ratio. The other disadvantage of this solution is that the set-up can only be used in machines which are able to follow the normal vector defined by the Tool Centre Point (TCP) and by the contact point of the sheet. The orientation change along the tool path is indispensable, because sheet thinning occurs in the small plastic zone defined in [20].

5 Novel On-Line Sheet Thickness Measurement Solutions in SPIF

In 2011, Paniti and Paróczi [15] introduced a different on-line solution for in-process thickness measurements which can be applied in SPIF. The authors analysed the possibility of using a Hall-effect sensor integrated into a flat incremental forming tool. The investigations in FEM modelling and simulation showed good results in the design phase regarding the measurement possibilities, but the flat forming tool design of the first prototype limited the orientation of the tool.

By keeping the measurement principle, an inverse set-up allows to measure the thickness in the small plastic zone. For the inverse set-up the forming tool suggestions of Leszak [12]—which have similarities to a ballpoint pen—gave a good starting point. The forming tool in this case is a bearing ball with free rotation possibility. To measure the thickness in SPIF with the same solution suggested in [15] a Hall-effect sensor and a magnet have to be placed on the other side of the sheet. This inverse set-up makes self-adjusting measurement possible, because the magnet follows the bearing ball along the programmed tool path and the orientation of the set-up is adjusted according to the thickness of the sheet in the small plastic zone.

5.1 Calibration of the Thickness Measurement Device

The measurement set-up consists of three main elements:

- a steel target ball (like a G100 CARBON type bearing ball with 10 mm diameter as used in our experiments);
- a linear-output Hall transducer (like an AD22151 type Hall sensor IC from Analog Devices in bipolar configuration [1] as used in our experiments);
- and a permanent magnet (like an AlNiCo permanent magnet with 10 mm length and 10 mm diameter as used in our experiments) that provides a magnetic field.

The magnetic flux density varies monotonically as a function of displacement, but the Hall voltage will be significantly nonlinear with respect to position. In this case the magnetic flux density is inversely proportional to the sheet thinning. For an accurate sheet thickness measurement it is important to define a calibration curve with discrete data acquisition. Figures 4 and 5 shows the calibration device with a micrometre screw as the carrier tool for the bearing ball.

Calibration measurements were carried out with 3 series (300 measurement points in each series) using a Motion Control and Data Acquisition PCI card [14] for the A/D conversion. Every 50 ms a 16-point moving average was applied as a low-pass filter. It is important to choose an adequate distance between the Hall sensor and the magnet, because a too high value could interfere with the motion of the device (see curve "magnet-sensor distance: 3 mm" in Fig. 6), while on the other hand a too small value could cause an overflow in the measurement (see curve "magnet-sensor distance: 1 mm" in Fig. 6).

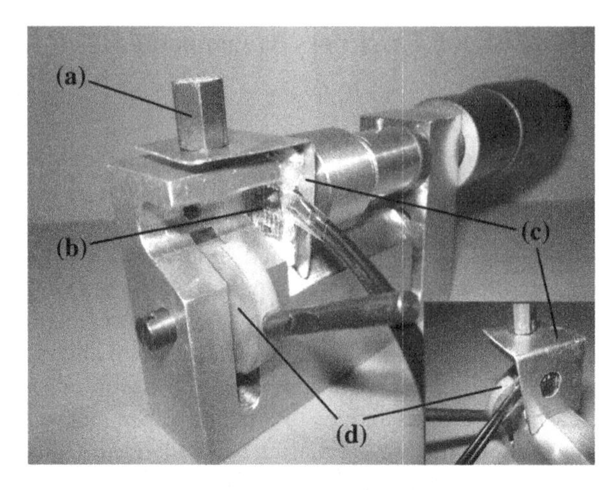

Fig. 4 Calibration set-up of the Hall sensor based thickness measurement device with **a** sheet mounting part, **b** Hall sensor, **c** bent Al sheet, **d** magnet mounting part

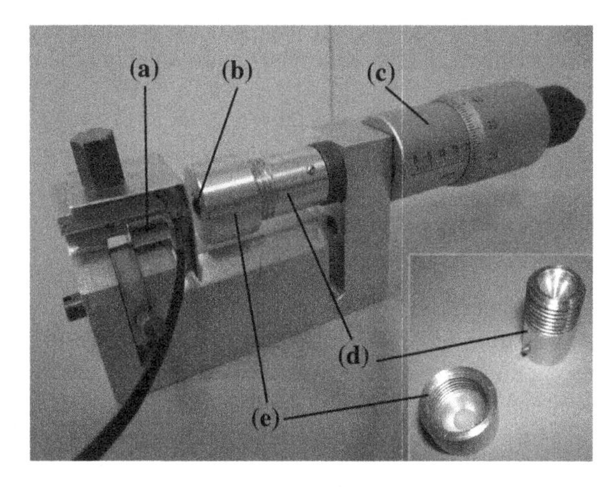

Fig. 5 Calibration set-up of the Hall sensor based thickness measurement device with **a** magnet, **b** bearing ball, **c** micrometre screw, **d** Al housing of the bearing ball, **e** cap of the Al housing

Figure 7 shows the calibration curve up to 1mm with an adequate magnet-sensor spacer glued between the AlNiCo magnet and the Hall sensor.

A 6th order polynomial regression in this case gives the following equation with a correlation value close to 1.

$$U = 26813t^6 - 12207t^5 + 19028t^4 - 12168t^3 + 2586.1t^2 - 244.32t + 3848.9 \quad (7)$$

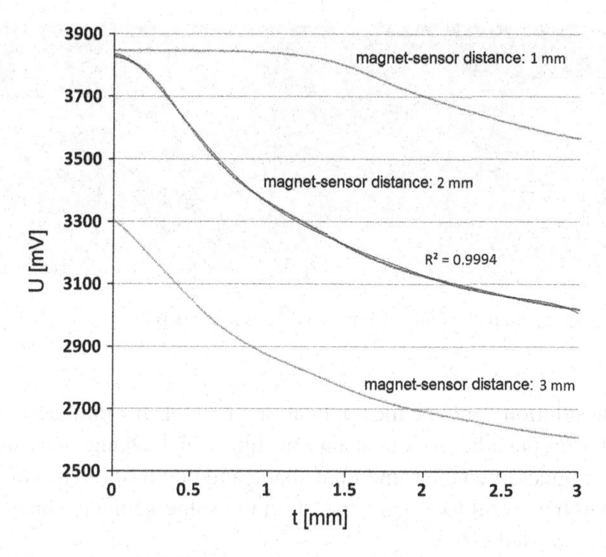

Fig. 6 Calibration curves with different magnet-sensor distances

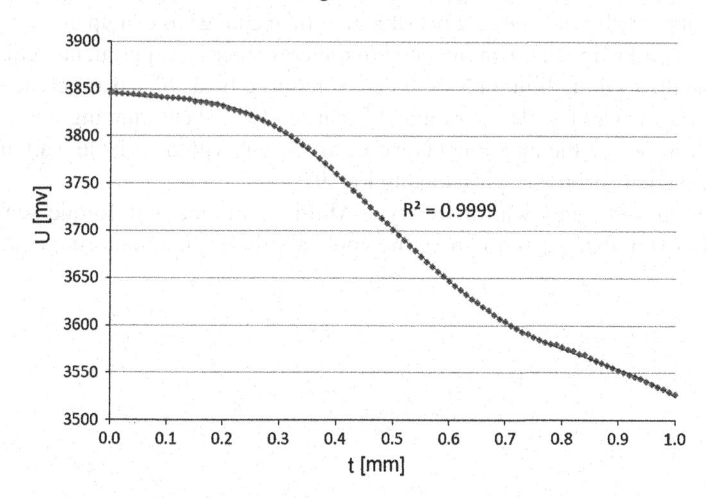

Fig. 7 Calibration curve of the prototype

5.2 Proof of Concept with Adaptive Control Algorithm

In 2004, Hirt et al. [10] summarised the influence of the SPIF process parameters on the final product and on each other. They concluded that sheet formability decreases by increasing the step depth.

In 2006, the experimental investigations of Attanasio et al. [4] showed the importance of using a tool path that has a variable step depth, because a lower step depth increases accuracy—but on the other hand it increases the process time too, thus leading to inefficient production.

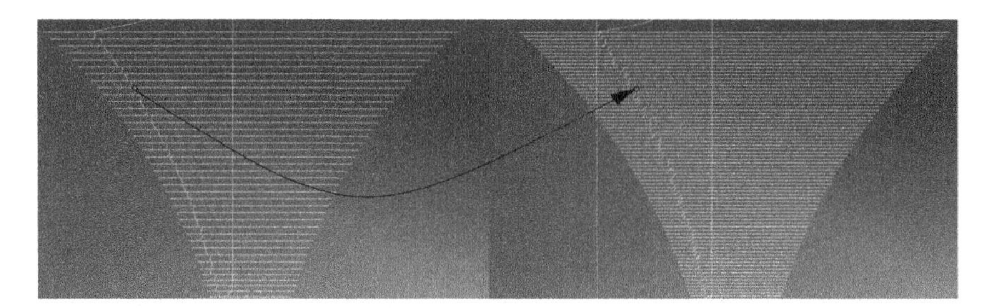

Fig. 8 Illustration of tool path change from 1 to 0.5 mm step depth

A possible solution could be the application of an on-line thickness measurement method with a simple adaptive control algorithm, which changes the step depth to a lower value. To guarantee the same final shape and depth of the product as with the initial step depth, several tool paths based on the same geometry but with different step depths are needed [16].

Figure 8 shows the two tool paths applied in the experiments with constant 1 and 0.5 mm step depths and shows a possible tool path change as example.

Hussain and Gao [11] demonstrated that the number of experiments required to determine the forming limit of a sheet can be reduced by using a part geometry with variable wall angle. For this reason and because of the sheet thinning observations discussed in Sect. 3 the experiments were carried out with a variable wall angle β, by using the part geometry presented in Fig. 9.

Force measurements with a JR3 type Multi-axis Force and Torque sensor are carried out simultaneously to prove the applicability of the monitoring concept by

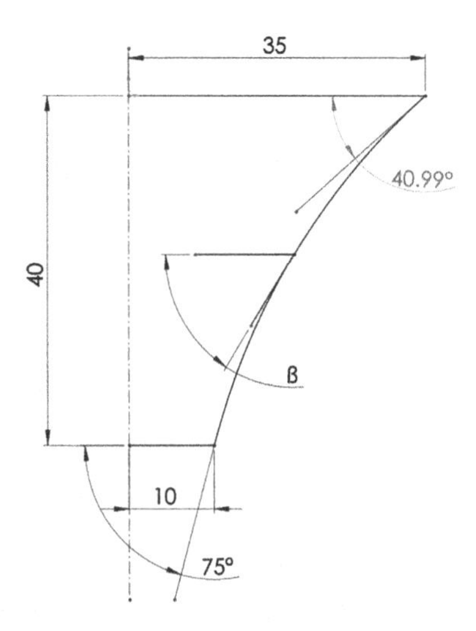

Fig. 9 Part geometry with variable wall angle

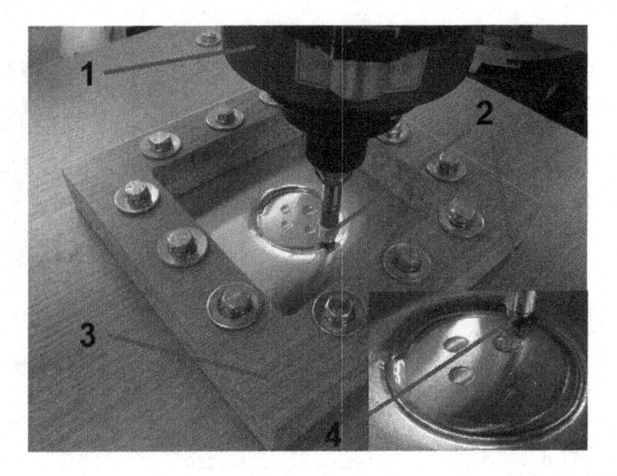

Fig. 10 SPIF set-up with *1* Force cell, *2* bearing ball (forming tool) holder, *3* clamping frame and *4* bearing ball as forming tool

Ambrogio et al. [3] on the part geometry. The force cell was mounted between the tool holder and the flange of a FANUC S430iF industrial robot.

In order to measure the sheet thickness with the Hall sensor based solution, but without disturbing materials in the close environment of the sensor, the clamping frame, the backing plate and the forming tool (bearing ball) holder in the set-up need to be non-ferromagnetic. Figure 10 shows the realisation of the previous mentioned criteria.

Machine oil was applied as lubricant on the surface of the sheet to minimise friction. Figure 11 shows the forming and the measurement set-up from below.

Fig. 11 SPIF set-up from below with *1* Hall sensor based thickness measurement device and *2* backing plate

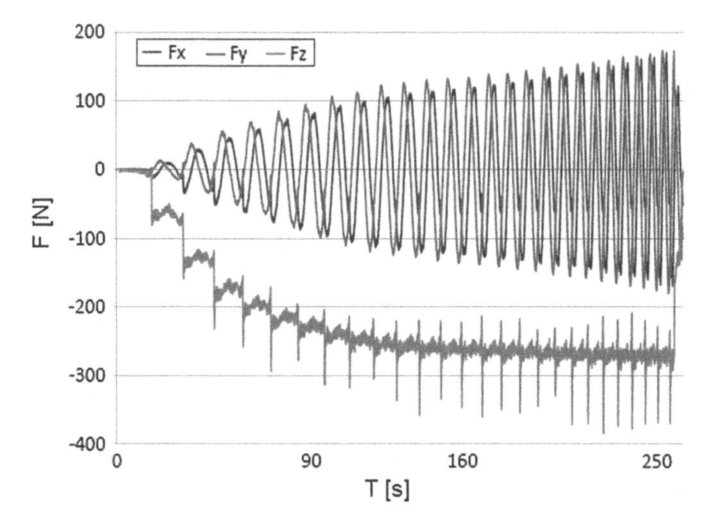

Fig. 12 Reaction force curves with sharp drop-down of the z-component (preliminary test)

Preliminary SPIF experiments were carried out with 0.5, 0.7 and 0.8 mm thick Al 1050 sheets and some experimental investigations were reported with 0.6 mm thick sheets in [23]. Despite the high wall angles in the part geometry, only those sheets with 0.5 mm initial thickness showed fracture in all cases. For this reason the target of the experiments was limited to the Al 1050 sheets with 0.5 mm initial thickness.

Due to the relative low magnetic density of the applied AlNiCo magnet the forming speed had to be limited to 9 mm/s to obtain reliable measurement results. Figure 12 shows a preliminary test with reaction force components measured by the Force cell till fracture.

Fig. 13 Photo of the formed part with circumferential crack propagation (preliminary test)

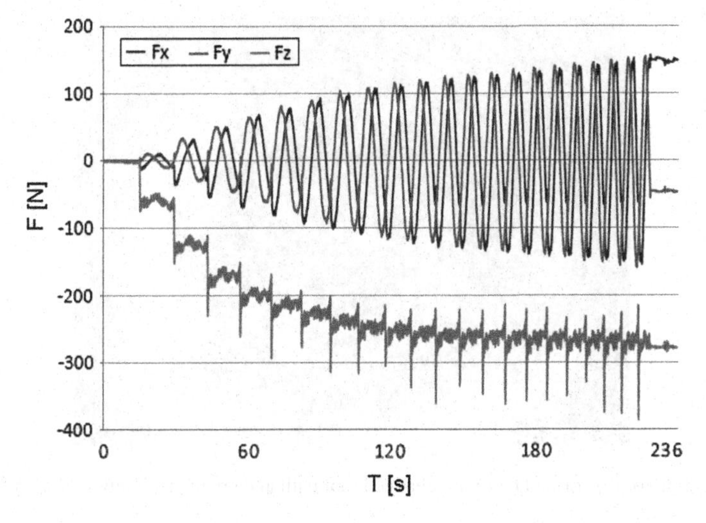

Fig. 14 Reaction force curves without significant change at the critical value (*1* test)

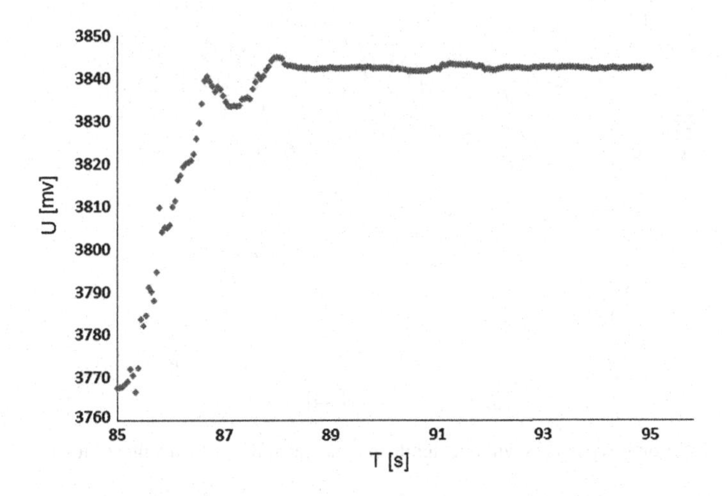

Fig. 15 Measurement results of the Hall sensor around the critical value (*1* test)

Fig. 16 Photo of the formed part with changed tool path and crack propagation (*1* test)

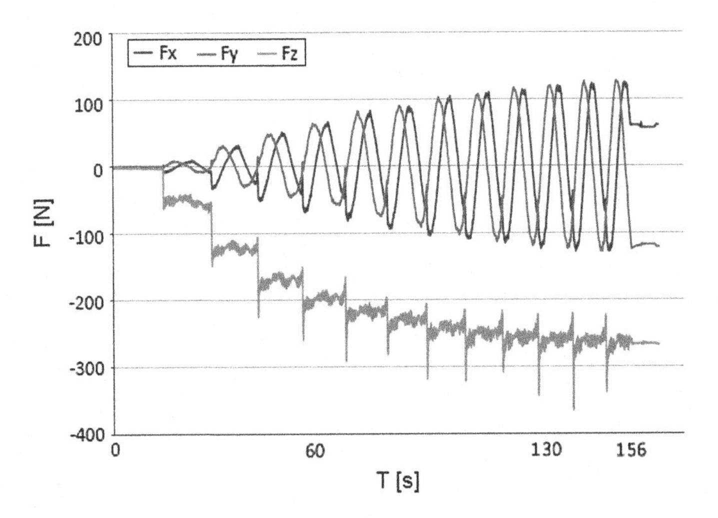

Fig. 17 Reaction force curves without significant change at the critical value (*2* test)

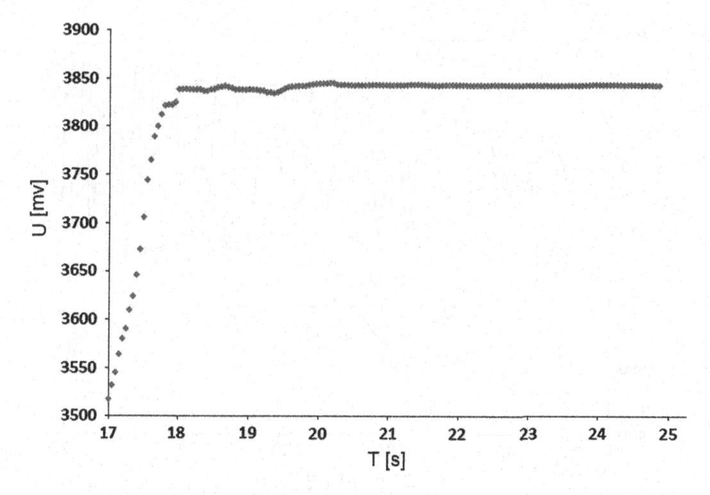

Fig. 18 Measurement results of the Hall sensor around the critical value (*2* test)

Fig. 19 Photo of the formed part with local thinning at the beginning of the forming (*2* test)

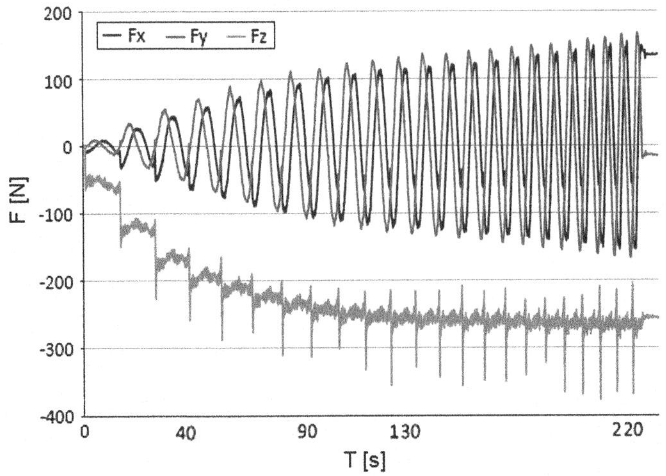

Fig. 20 Reaction force curves without significant change at the critical value (*3* test)

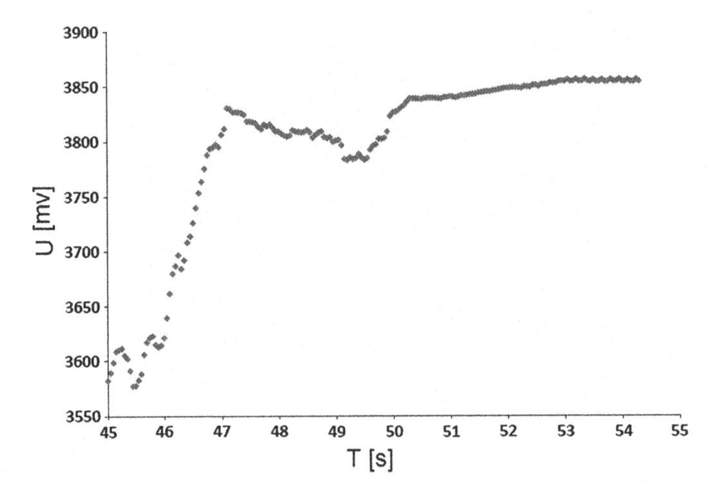

Fig. 21 Measurement results of the Hall sensor around the critical value with a "stabilised thickness area" (*3* test)

Fig. 22 Photo of the formed part with changed tool path and crack propagation (*3* test)

A clear drop in the z-component (Fz) could be observed at the time of the fracture. The result of the forming on the part is shown by Fig. 13.

Hall voltage around 3840 mV was measured at the fracture point. This corresponds to a thickness of 0.11 mm ± 0.01mm. The thickness measurement was carried out after the forming to determine the critical value at the fracture point. Figures 14, 17 and 20 shows the reaction force components (for the same part) up to the critical value obtained by the Hall sensor based thickness measurement device before the tool path change.

The current prototype size of the Hall sensor based thickness measurement device and the backing plate did not allow the measurements to be made from the beginning. Results (Figs. 15, 18, 21) show the last 150–200 measurement points around the critical value for the same part, where the forming was stopped.

Figure 16 shows the formed part with 1 and 0.5 mm step depths. Crack occurred after the tool path change.

Figure 19 shows that a local thinning occurred with a critical value at the beginning of the forming. The process was stopped according to the algorithm and a tool path change was started. The tool path with 0.5 mm step depth formed the part almost completely.

Figure 21 shows a "stabilised thickness area" like in [25] without continuous thinning. The critical value is reached twice in this case. Figure 22 shows the formed part of the third test with changed tool path and crack propagation.

This thickness measurement solution is limited to non-ferromagnetic materials, but if the conditions are adequate (technical properties are selected properly and the device is secured against flipping to the lower flat zone of the sheet with e.g. a solution such as that reported in [22]) mean measurement accuracy equal to ±0.01 mm with proper repeatability can be reached.

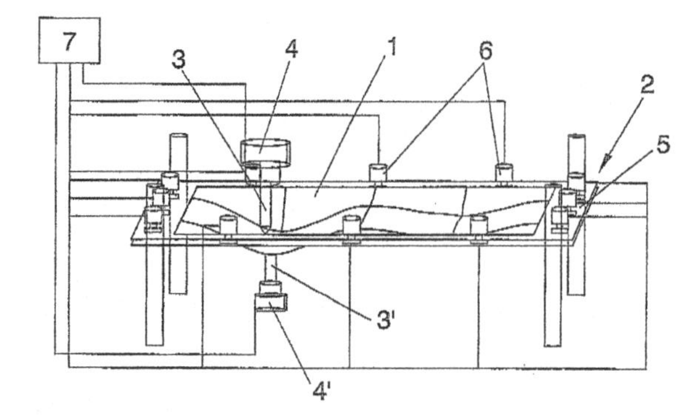

Fig. 23 Patented ISF solution of Rodriguez Gutierrez et al. [19] with *1* sheet, *2* fastening, *3* first tool, 3': second tool, *4* mounting head of first tool, 4': mounting head of second tool, *5* peripheral frame, *6* piezoelectric actuators, *7* controller

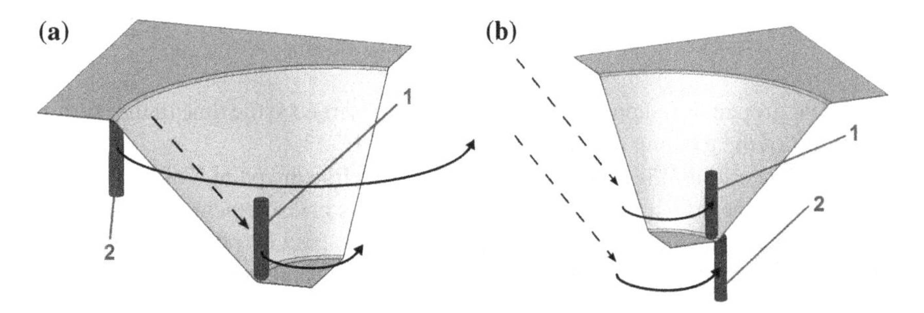

Fig. 24 **a** Forming strategy "A", **b** Forming strategy "B" where *1* first forming tool, *2* second forming tool

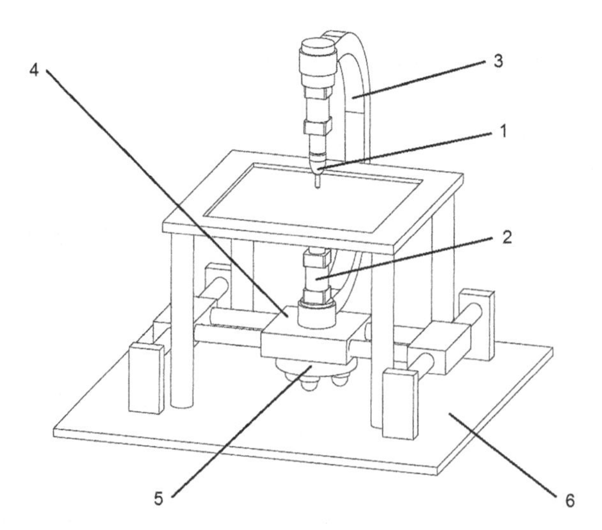

Fig. 25 Patented ISF solution of Somlo and Paniti [22] where *1* first forming tool, *2* second forming tool with linear actuator, *3* C-frame, *4* linear slide system, *5* ball transfer unit, *6* base plate

6 Novel TPIF Solutions with Counter Tool

In the previous years an enhancement in ISF could be observed which manifested itself in a number of R&D projects and patents [8] all over the world.

SPIF has several limitations; for example it needs a backing plate, and forming of convex and concave parts on the same sheet is only feasible if the clamping frame is released.

The patented solution of Rodriguez Gutierrez et al. [19] overcame these issues by the synchronised movement of two forming tools (see Fig. 23). The synchronisation of the two tools is solved by a common controller.

Forming experiments on the prototype of the previously mentioned patent showed that the set-up allows two main forming strategies (see Fig. 24) for the manufacturing of sheet metal parts and the local support of the second tool gives better results in terms of formability [23]. Experiments were carried out on a 0.6 mm thick aluminium sheet (Al 1050).

In the case of strategy "A" the second tool acts as a backing plate, moving with the first tool, but does not leave the starting contour of the supporting tool path. In the case of strategy "B" the second tool moves with the forming tool, ensuring continuous support. The same solution with two synchronised industrial robots was used by Meier et al. [13]. They also observed better results with strategy "B"; their names for the abovementioned strategies were "duplex incremental forming with peripheral support" (DPIF-P) and "duplex incremental forming with local support" (DPIF-L) respectively.

Meier et al. [13] extended the analytical model of Silva et al. [20] for rotational symmetric SPIF by the superimposed pressure induced by the second (supporting) tool to analyse the influence of the parameters in DPIF-L.

Similar results can be reached with the patented solution of Somlo and Paniti [22]. The device also makes it possible to form a sheet from both sides, but instead of the synchronisation of a minimum 2×3 axis the x-y motion of the second forming tool is copied with a C-frame (see a variant of this in Fig. 25).

In Fig. 25 a variant of the patented device is illustrated with a linear slide system which ensures a smooth motion in the x-y direction. The axes of the linear slide system are secured against deflection by a ball transfer unit containing a minimum of one iron ball. The ball transfer unit is placed on a smooth base plate.

This device can be used as an external axis in some milling machine centres and in robotic cells. A detailed description of the solution can be found in [22].

7 Conclusions

In this study we intended to show the importance of an adequate sheet thickness measurement solution due to the limitations of thickness prediction approaches.

The presented results showed that the Hall sensor based thickness measurement method for SPIF is capable of measuring the thickness in the small plastic zone of the sheet and can detect stabilised thickness areas. The results verified that the solution can be coupled with the adaptive control algorithm presented in [16]. Experiments showed that in the examined cases the monitoring of the pre-defined critical value was able to indicate the need for a tool path change. Fracture occurred only after the execution of the tool path with smaller step depth. Furthermore some novel solutions were introduced in TPIF / DPIF (duplex incremental forming).

Acknowledgments The described work was carried out as part of the TÁMOP-4.2.2/B-10/1-2010-0008 project in the framework of the New Hungarian Development Plan. The realization of this project is supported by the European Union, co-financed by the European Social Fund. The author would like to also thank the Institute for Computer Science and Control, Hungarian Academy of Sciences, for its support.

References

1. AD22151 type Hall sensor IC. http://www.analog.com/static/imported-files/data_sheets/AD22151.pdf
2. Aerens, R., Eyckens, P., Van Bael, A., Duflou, J.: Force prediction for single point incremental forming deduced from experimental and FEM observations. Int. J. Adv. Manuf. Technol. **46**, 969–982 (2010)
3. Ambrogio, G., Filice, L., Micari, F.: A force measuring based strategy for failure prevention in incremental forming. J. Mater. Process. Technol. **177**(1–3), 413–416 (2006)
4. Attanasio, A., Ceretti, E., Giardini, C.: Optimization of tool path in two points incremental forming. J. Mater. Process. Technol. **177**(1–3), 409–412 (2006)
5. Avitzur, B., Yang, C.T.: Analysis of power spinning of cones, J. Eng. Ind.-Trans. ASME, B **82**, 231–245 (1960)
6. Bambach, M.: A geometrical model of the kinematics of incremental sheet forming for the prediction of membrane strains and sheet thickness. J. Mater. Process. Technol. **210**(12), 1562–1573 (2010)
7. Dejardin, S., Gelin, J.-C., Thibaud, S.: On-line thickness measurement in incremental sheet forming process. Proceedings of the 13th international conference on metal forming, Toyohashi, 19–22 Sept 2010, pp. 938–941
8. Emmens, W.C., Sebastiani, G., van den Boogaard, A.H.: The technology of incremental sheet forming—a brief review of the history. J. Mater. Process. Technol. **210**(8), 981–997 (2010). doi:10.1016/j.jmatprotec.2010.02.014
9. He, S., Gu, J., Sol, H., Van Bael, A., van Houtte, P., Tunckol, Y., Duflou, J.R.: Determination of strain in incremental sheet forming process. Key Eng. Mater. **344**, 503–510 (2007). doi:10.4028/www.scientific.net/KEM.344.503
10. Hirt, G., Ames, J., Bambach, M., Kopp, R.: Modeling and experimental evaluation of the incremental CNC sheet metal forming process. CIRP Ann. Manuf. Technol. **53**(1), 203–206 (2004)
11. Hussain, G., Gao, L.: A novel method to test the thinning limit of sheet metal in negative incremental forming. Int. J. Mach. Tool. Manuf. **47**, 419–435 (2007)
12. Leszak, E.: Apparatus and process for incremental dieless forming, US Patent 3342051A1 (1967)
13. Meier, H., Magnus, C., Smukala, V.: Impact of superimposed pressure on dieless incremental sheet metal forming with two moving tools. CIRP Ann. Manuf. Technol. **60**(1), 327–330 (2011)
14. Datasheet of the MOTENC-Lite PCI card. http://www.vitalsystem.com/web/motion/motionLite.php

15. Paniti, I., Paroczi, A.: Design and modeling of integrated Hall-effect sensor based on-line thickness measurement device for incremental sheet forming processes. In: Proceedings of the 2011 IEEE/ASME International Conference on Advanced Intelligent Mechatronics (AIM), http://ieeexplore.ieee.org/xpl/articleDetails.jsp?arnumber=6027146pp. 297–302. Budapest, Hungary, (2011)
16. Paniti, I., Rauschecker, U.: Integration of incremental sheet forming with an adaptive control into cloud manufacturing. In: Proceedings of the 14th International Conference on Modern Information Technology in the Innovation Processes of Industrial Enterprises, MITIP2012, pp. 119–128. Budapest, 24–26 Oct (2012)
17. Powell, N., Andrew, C.: Incremental forming of flanged sheet metal components without dedicated dies. IMECHE part B. J. Eng. Manuf. **206**, 41–47 (1992)
18. Rauch, M., Hascoet, J.-Y., Hamann, J.-C., Plenel, Y.: Tool path programming optimization for incremental sheet forming applications. CAD Comput. Aided Des. **41**(12), 877–885 (2009)
19. Rodriguez Gutierrez, P.P., Rivero Rastrero, M.A., Maidagan Onandia, E.: Machine for shaping sheet metal and shaping method, EU Patent EP1977842 (2006)
20. Silva, M.B., Skjoedt, M., Martins, P.A.F., Bay, N.: Single-point incremental forming and formability-failure diagrams. J. Strain Anal. Eng. Des. **43**(1), 15–36 (2008)
21. Silva, M.B., Skjoedt, M., Bay, N., Martins, P.A.F.: Revisiting single-point incremental forming and formability/failure diagrams by means of finite elements and experimentation. J. Strain Anal. Eng. Des. **44**, 221–234 (2009). doi:10.1243/03093247JSA522
22. Somlo, J., Paniti, I.: Device for two sided incremental sheet forming. EU Patent EP2505279 (2012)
23. Tisza, M., Paniti, I., Kovács, P.Z.: Experimental and numerical study of a milling machine-based dieless incremental sheet forming. Int. J. Mater. Form. **3**(1), 441–446 (2010)
24. Van Bael, A., He, S., Van Houtte, P., Tunkol, Y., Verbert, J., Duflou, J.R.: Study on the thinning during single point incremental forming of aluminium sheets. Presented at the 24th international deep-drawing research group congress, p. 12. Besançon, France (2005) http://www.ens2m.fr/iddrg2005
25. Young, D., Jeswiet, J.: Wall thickness variations in single-point incremental forming. J. Eng. Manuf. Part B **218**, 1453–1459 (2004)

New Online Flicker Measuring Method and Module

Attila Unhauzer

Abstract Numerous non-linear equipment are connected to the electrical energy distribution system and increasing continuously. Today, the reduction of disturbance effects is of great significance since entirely new regulations, standards and requirements have been introduced by the European Union. These standards extend to the measurement methods of electrical networks but also to the limits of the most important electrical parameters. In this chapter, a new online flicker measuring system is presented which is suitable for measuring the relevant flicker parameters according to the current standards of energy quality in a more flexible way. The developed new flickermeter was also built in a modern network diagnostics system accomplishing more real measurement by transformer stations of power companies. This meter and connected industrial equipment were scrutinised using a new modern calibration method.

1 Introduction

High-quality electrical energy serves the interests of both consumers and suppliers. The limits of electrical network parameters are increasingly important nowadays, since they are not only an essential component of adequate, continuous, secure and economical operation, but in certain cases are crucial for human health as well as with the flicker phenomenon.

Flicker is a sequence of flashing lamp pulses which imperceptibly influences the human body and the environment. It is caused by fast voltage fluctuation and is usually produced by non-linear high-capacity equipment, e.g. switching effects of electric

A. Unhauzer (✉)
Department of Electrical and Electronic Engineering, University of Miskolc,
Miskolc-Egyetemváros 3515, Hungary
e-mail: unhauzer@gmail.com

G. Bognár and T. Tóth (eds.), *Applied Information Science, Engineering and Technology*, Topics in Intelligent Engineering and Informatics 7, DOI: 10.1007/978-3-319-01919-2_10,

arc furnaces, high-power loads, starting up high-power motors and high-performance welding.

This chapter describes a new online flicker measurement method and a multiple-tested software module based on modern multithreading technology that have been developed for the objective and exact analysis of electrical networks. The necessary theoretical parts and the main development steps will be covered. Full development and further description can be found in [1].

2 Scientific Background

A great number of digital but few online flicker measuring methods are available; however, most existing methods and algorithms are simulations only. A good range of flicker measurement equipment can be purchased on the market; the computation method for such equipment is not published by the manufacturer. Most developed modules cannot be installed into a reliable and flexible measuring system.

In addition, the existing measurements are not able to provide enough flexibility concerning the parameters of flicker measurement, although flexibility is crucial in the design of a universal system. Each possible parameter modification, such as the setting of sampling rate, alterability and the fine-tuning of the flicker perception curve, is covered in the technical literature. The adaptability of the sampling rate is extremely important, especially in the case of phase jumps, which have been discussed only by some theoretical (e.g. [2]) and even fewer practical research papers (e.g. [3]). The alterability and the fine-tuning of the flicker perception curve plays an important role due to the large number of new light sources available today, such as energy-saving and LED lights [4]. Many research publications cover the necessity of the objective measurement method [5], which is essential to compare flicker values independent of types of lamps [6].

The architecture and calibration of flicker measurements are given in the standards [7–9]; there remain, however, numerous opportunities for implementation and testing. Most flickermeters are based on methods of analyses such as Fourier [10], RMS[1] [11], peak value [12] and wavelet analyses [13]. Digital methods look to be widespread today.

In order to emphasize the complexity of the methods, most research output focuses on the easy solutions of flicker measurements. The objective comparability of flicker measurement systems [14] plays a central role in several studies, including studies of automatic and manual calibration [15], error correction [16], error measurements [17], and sensitivity of measurements [18].

[1] Root Mean Square.

3 Flicker Phenomenon and Standardised Measurement

3.1 Human Sensitivity According to Flicker Perception Curve

It is difficult to measure objectively the disturbing effect of flashing, and one standard [7] therefore describes the perception process of voltage fluctuation lamp-eye-brain response, which precisely indicates the reaction of the observed person to flashing. On the basis of observations of human beings and animals, it has been found that humans are particularly sensitive to voltage fluctuation of around 8.8 Hz. Figure 1 shows the relative voltage changes as a function of the number of rectangular voltage changes per minute where 1,056 voltage changes per minute are relevant to 8.8 Hz in the frequency domain. Consequently, the standards stipulate that mains voltage must not contain these frequency components.

3.2 Flicker Measurement Suggested by the Standards

The latest standard [9] defines the analogue architecture of the flicker-meter, determining the components of flicker measurement by describing each functional block. The architecture can be divided into two functional elements (Fig. 2) as follows:

- The flicker perception of human beings and lamp-eye-brain response are simulated by the 2nd, 3rd and 4th cascade of blocks.
- The *Block 5* represents the online statistical analysis module; it computes the short and long-term flicker severity index (P_{st}, P_{lt}).

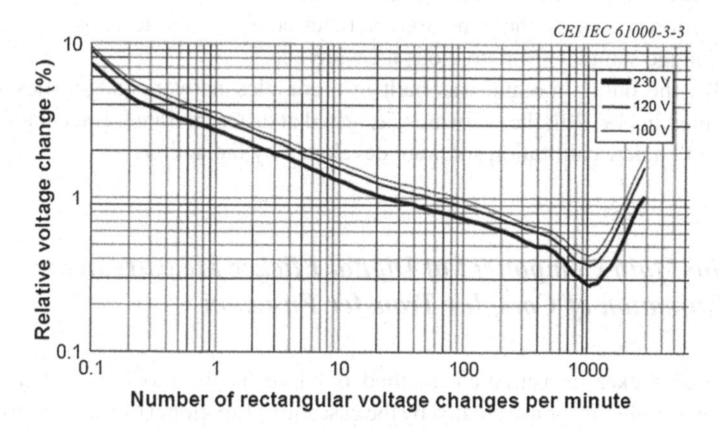

Fig. 1 Flicker perception curve

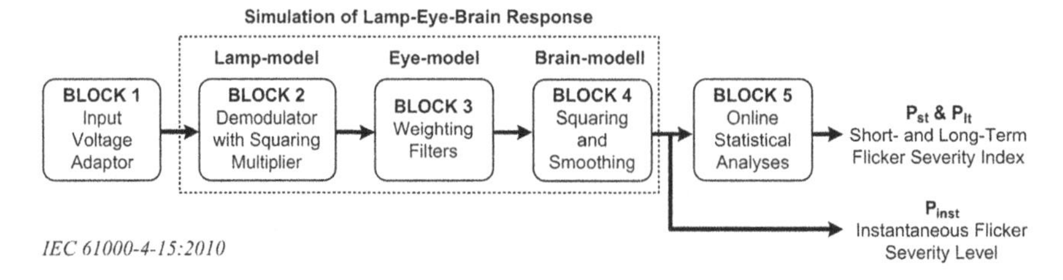

Fig. 2 Simplified block diagram of a flickermeter

4 Modern Online Method for Flicker Measurement

4.1 Introduction to the Project

The method of flicker measurement is also determined, by other standards [7–9]. The flickermeter module analyses the voltage signal through a complex mathematical transformation process. The development process of this work was divided into three main parts for simplicity:

- In the first phase, the modular blocks of flickermeter (simulation of lamp-eye-brain blocks and statistical block) were developed to be able to simplify the creation processes. Each flickermeter block satisfies the accuracy and working speed performing the standard requirements. The development of central block meant the main step during the work which supports the simulation of eye-model by digital filtering.
- After the development process of modular blocks, a new online flickermeter was developed using modern multithreading technology for data estimation and communication processes; these methods were tested with a new test application using real voltage samples from previous measurements.
- Finally, the online module was built in a complex network diagnostics system described in [19, 20]; the software was tested at different transformer stations and also with a new calibration software developed by the author.

4.2 The Main Component of Digital Flicker Measurement: Estimation of Complex Transfer Function

The digital flicker measurement method is based on the modelling of nonlinear *"lamp-eye-brain"* response created by the cascade of functional blocks. More blocks, e.g. *Block 2* and *4*, can be simulated with simple mathematical signal transformation steps, e.g. squaring, multiplication and easily tuneable IIR[2] Butterworth filtering processes following the analogue description of the standard. Hence, only the analogue measurement method (system) is defined by the standard and the digital transformations are found only in few cases.

During digital implementation, the approximation of human eye sensitivities caused the major problem. Eye sensitivity can be characterized by the complex transfer function in Eq. (1) with the behaviour of continuous time domain. The main task is to transform the *A(s)* formula into a discrete time domain for digital signal processing.

$$A(s) = \frac{k\omega_1 s}{s^2 + 2\lambda s + \omega_1^2} \frac{1 + \frac{s}{\omega_2}}{(1 + \frac{s}{\omega_3})(1 + \frac{s}{\omega_4})} \tag{1}$$

Block 3 (cf. Fig. 2) is a crucial component of the digital flicker measurement method and its accurate approximation is the most important step in the development project.

Main Steps of Development. This research began in several directions simultaneously to approximate the eye model. During this process, the accuracy and processing speed factors were also observed, following the requirements of an online system.

Thus, the complex transfer function was approximated applying IIR filters (digital filter cascade), Fourier-transform and z-transform while checking the process speed and the accuracy of transformations. Finally, with the help of bilinear discrete z-transform (Tustin formula), a digital IIR filter was created which gave the optimal result with minimal error, high speed, and reliable functioning.

Digital Filter Design Using the Tustin Formula. The advantage of the bilinear z-transform is stability; thus, the stability criterion is fully satisfied by the filter. During the transformation process, the domain s was transformed into domain z (Eq. 2), applying the Tustin formula in Eq. (3). The well-known types of filters, e.g. Butterworth, elliptic and Chebychev, can also be simply transformed by using this formula:

$$z = e^{sT_0} \tag{2}$$

$$s = \frac{1}{T_0} \cdot \ln z \approx \frac{2}{T_0} \cdot \frac{z-1}{z+1} \tag{3}$$

In the Eqs. (2) and (3), s is the Laplace operator (complex frequency domain); z is the operator of z-domain (discrete time domain); z^{-1} indicates the delaying process of this transform, which means an earlier computed variable; and T_0 is the sampling time.

During the transformation process, the Tustin formula is substituted into the complex transfer function of Eq. (1), which results in a hugely unordered equation. After the ordering and simplifying processes of this equation, the digital filter coefficients will also be determined, resulting in the required digital filter. The high complexity of the digital filter can be easily perceived with the transversal model (Fig. 3).

[2] Infinite Impulse Response.

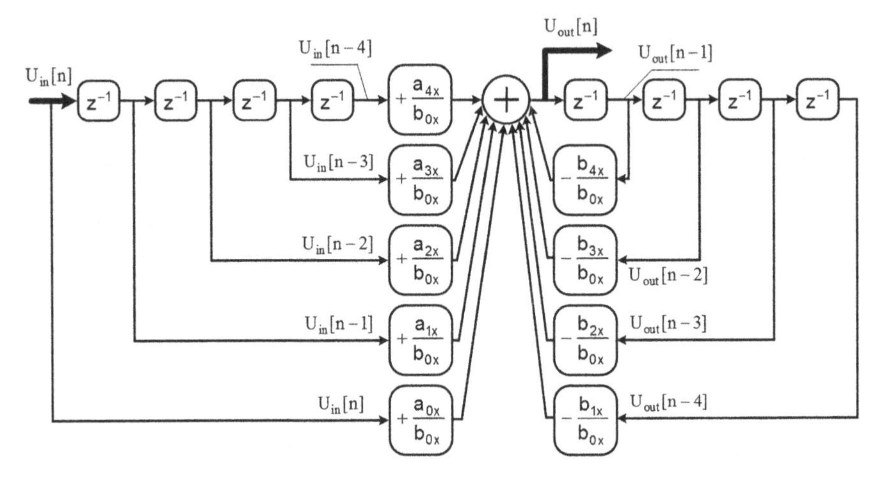

Fig. 3 The transversal model of digital filter, indicating its coefficients

Testing of the New Digital Filter. Simulation software was developed for testing the accuracy and correct functioning of the digital filter which generates specialized signals for testing. This software works with high precision ($10^{-12} \cdot 10^{-16}$) which is required to be reliable in order to enable the comparison of the results with the standard [9].

According to the standard [9], the developed new filter had to be tested under 35 Hz frequency, because the human body's sensitivity is influenced by this significant frequency spectra. For filter testing, the special input signal in Eq. (4) was generated by the simulation software, which has 35 different frequency harmonics (1–35 Hz) with unity amplitude. The selected frequency domain and unity amplitude of the frequency components both suffer from the fact that the transfer function is given immediately after the input signal is filtered (Fig. 4). The simulation software displays (Fig. 4) the spectra of *A(s)* (1) complex function (enveloped curve) according to standard [9], the spectra of the developed filter (vertical lines) and a relative error curve between these two transfer functions.

$$u_{generated}(t) = \sum_{j=1}^{35} U_{\max} \cdot \sin(\omega_j \cdot t) \text{ where } U_{\max} = 1\,[V] \tag{4}$$

In the equation $u_{generated}(t)$ is the complex, 60 s time signal, U_{max} is the amplitude of generated signal and ω_j is the cyclical frequency of generated signal.

The filter tests also cover different sampling rates. The operation of the digital filter is not influenced by the changing sampling frequency, which is very important for general flexibility. The filter error is less than 2.3 % even at or under frequency 5 Hz; thus, it is adequate for later adaptability and applicability. Consequently, the

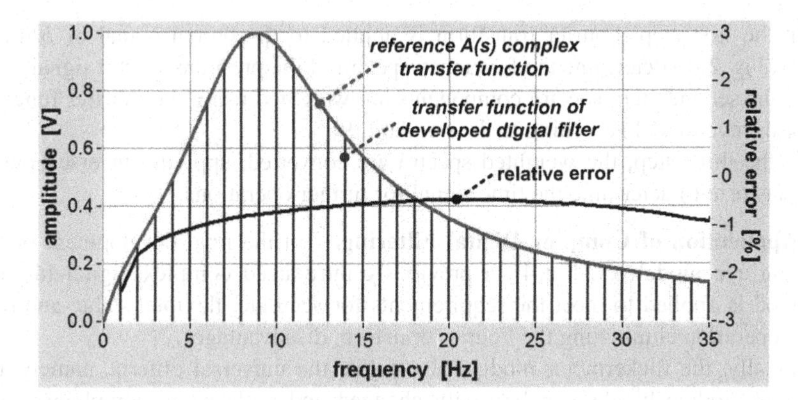

Fig. 4 Transfer function of the developed digital filter (*vertical lines*) compared with the reference $A(s)$ complex transfer function (*envelope curve*), showing the relative error between the two functions

created filter gives an optimal solution; it shows a short operation time, high accuracy and flexibility.

According to the 2010 version of the standard requirements [9], the flickermeter error has to be less than 5 %. The approximation of the complex transfer function (Eq. 1) has central importance, since most flickermeter components are simply mathematical transformations, whereas the human eye model is a sophisticated formula and difficult to approximate. Consequently, the complex digital filtering process determines the accuracy of the flickermeter module. According to my experience, the approximation of the digital filter is accomplished successfully.

4.3 Testing of the Digital Flicker Measurement

Simulation software was developed for the technological development of the flicker measurement method realizing every transformation block of flickermeter. During the implementation, the statistical block (*Block 5*, cf. Fig. 2) is also developed, applying the time-at-level method.

Two different testing methods and software are developed, which are different only in the realization of the eye model (*Block 3*, cf. Fig. 2). A combination of Fourier and inverse Fourier transforms as well as complex digital filtering are also used for the realization of the eye-model transforming samples.

New Method Using Fourier and Inverse Fourier Transforms. This method in particular allowed the development of every transformation block and the creation of a reliable, stable signal transformation process for determining the short and long-term flicker level (P_{st}, P_{lt}). Human eye modelling is developed with three-step transformation:

- In the first step, Fourier transform is applied to the output signal of *Block 2* (cf. Fig. 2) to determine the frequency spectra of the pre-transformed signal.
- In the second step, spectra components are weighted using the transfer function of the eye model according to the standard [9].
- In the third step, the weighted spectra are converted, applying inverse Fourier transform to determine the time signal for further operations.

Application of Complex Digital Filtering. In the final development of the human eye model (cf. Fig. 4), a previously introduced complex digital filtering method is applied to meet the requirements for accurate, flexible, stable and reliable operation eliminating the Fourier transform disadvantages.

Finally, the flickermeter module also fulfils the universal criteria, namely, the applied module blocks can be easily changed and replaced in compliance with the latest standards. The modular architecture and easy component adaptation of the flicker method also facilitate further modification, which is important given the fast-changing energy quality requirements of today.

4.4 Online Flickermeter Module

After developing the units of flickermeter, including the development of cascade of blocks, the background environment of online working was realized. The compact interfaces, e.g. few input and output connection points applying simple data communication and universal solutions, e.g. increasable sampling rate, were also used for further limitations of standard requirements.

Communication Interfaces of the Flickermeter Module. The main task of the online flickermeter module is to continuously compute P_{st} (at 1, 2, 5, 10 min) and P_{lt} values, saving all the data simultaneously. The simplest interfaces assisting further connection possibilities to other systems were defined. Consequently, the online module communicates through four compact connection points (Fig. 5). The module has a data acquisition input for voltage samples, and an input for sampling rate settings which has to set according to the standard [9] using a minimum of 450 Hz. It has a computation output for short-term flicker values and an optional output defined for the instantaneous flicker severity level, which gives a directly applicable (transformed) time signal for the purpose of statistics.

The developed module can be regarded as a *black box*, containing all the operational principles. In order to use the flickermeter module, it is sufficient to know

Fig. 5 Interfaces of the online flickermeter module

only its functional aims and interface parameters, because it is an autonomic unit. The flicker parameter is computed by the modern multitasking method.

Flexible Adjustable Flickermeter Module. The applicability of the online flickermeter module is supported first of all by the easily adjustable sampling rate. The appropriate setting of the module's sampling rate is also dependent on the standards in effect and on further measurement requirements, e.g. Shannon's theorem. In addition, the latest standard requirements [9] define three new flickermeter *Classes* F_1, F_2, and F_3 which require the changing of the sampling rate.

Class F_1 flickermeters are created for general purpose which are suitable for power quality monitoring as well as compliance testing; *Class* F_2 flickermeters intended for product compliance testing according to standard [7] in a controlled environment, with constant frequency and phase, and limited voltage fluctuations (*Class* F_2 is a specialized calibration category); finally, *Class* F_3 flickermeters are intended for use in power quality surveys, trouble shooting and other applications where low measurement uncertainties are not required.

In today's modern multitasking programming, only one data acquisition card is used for multichannel sampling. In addition, every electrical quality parameter, e.g. power, reactive power, RMS, THD,[3] flicker and so on, will be computed online where different settings for the sample rate are needed by different tasks. The mentioned problem is solved by modern software that defines only one data acquisition task using a high sample rate with a special FIFO[4] memory; the further tasks, independently of the sample rate, can access the data samples from FIFO in order to compute the parameters.

The general tendency is to minimize each computational demand during the online data estimation processes in order to be able to decrease the processor load and compute more parameters simultaneously. Therefore, the tasks have to achieve the parameter requirements with a minimal number of sample rates. In most cases, anti-aliasing filtering and down-sampling methods are used for decreasing the sampled data [21]. In this way, larger amounts of data and the sample rate can both be efficiently decreased (down-sampling method) while paying special attention to the requirements of frequency analyses, e.g. the last harmonics for analysis in the spectra. Beside down-sampling, using a high-order low-pass filter (anti-aliasing filter) guarantees the elimination of unwanted frequency components from the spectra (Fig. 6). Realization of down-sampling data and anti-aliasing filtering requires great attention, above all for FIFO-based operations.

Modern Multitasking Method for Flicker Measurement. During the online operation, the following tasks have to be solved (Fig. 7):

- data acquisition has to be performed on the selected input channels;
- down-sampling and anti-aliasing filtering processes have to be performed and synchronized with multichannel data acquisition;
- after the observation periods (e.g. 1, 5, and/or 10 min) estimation of short-term flicker level (P_{st}) has to be evaluated.

[3] Total Harmonic Distortion.

[4] First-In-First-Out Memory.

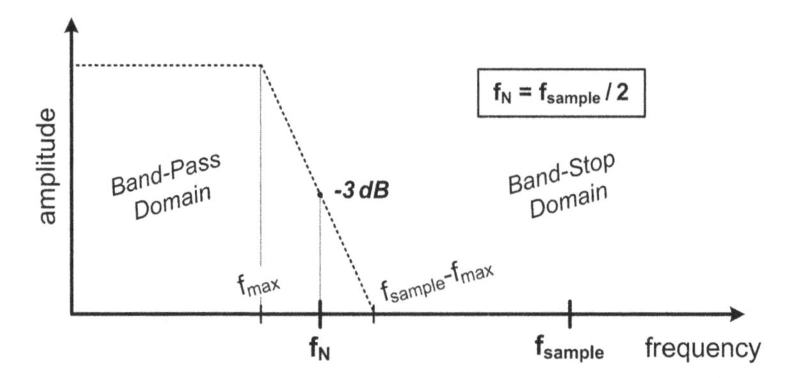

Fig. 6 Correct cut-off frequency settings for an anti-aliasing filter

Fig. 7 Functional architecture of the online flickermeter module

All of the data sampling, down-sampling and anti-aliasing processes demand highly synchronized operation; the estimation of flicker values is an asynchronous task. For the synchronized operations, an optimal-sized FIFO memory is needed to help the coordination of writing and reading samples.

The evaluation of flicker values is solved using further memory areas. If the sample rate of central data acquisition is the same as the rate of the flickermeter module then down-sampling and anti-aliasing can be left to special cases. In general, the determination of optimal sample rate is needed for the correct operation of the online module, which mainly depends on the one hand on standard requirements and on the other hand on certain technical requirements (e.g. calibration).

Figure 8 presents the architecture of the online module, which is connected to a data acquisition task operating with a high sample rate. The modular architecture shows the FIFO memory of Thread 1 supporting the asynchronous data communication between two different threads, i.e., between the acquisition and flicker estimation threads. The data communication through FIFO is based on the optimal-sized packages (sampled data from different channels) ensuring asynchronous data flow handling.

In this architecture, *Thread 1* ensures the data acquisition and communication, while *Thread 2* handles the down-sampling and anti-aliasing processes, expanding the advanced buffering solution in order to estimate flicker level. The buffer temporarily stores the down-sampled and anti-aliased samples before the evaluation of short-term flicker values.

A multiphase method is developed for the estimation of flicker level (*Thread 2*) which efficiently handles the temporary data buffering and computation processes

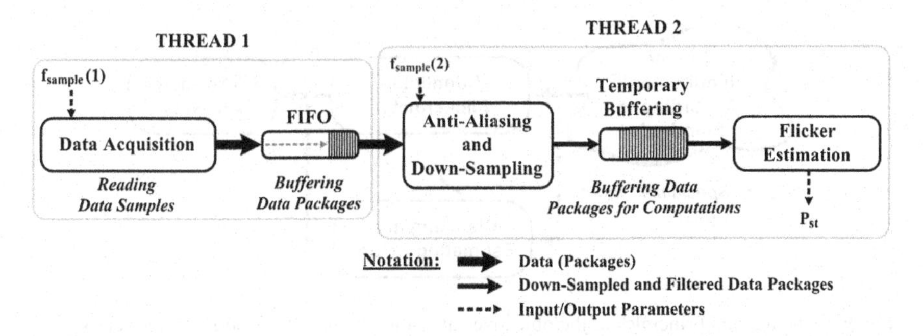

Fig. 8 Data communication architecture of the new online flickermeter module

in contrast to traditional brute-force estimation methods. More efficient methods are required when applying a higher sampling rate and short-term flicker values larger than 1 min.

The problem concerned is that the signal transformation process of flicker computation cannot be divided into smaller tasks on the one hand and pre-processing is not possible on the other. For example, for a multichannel (3, 6, 9, 12 etc.) online flicker estimation applying a 5–10 kHz sampling rate, $P_{st\ 5\ \min}$ and $P_{st\ 10\ \min}$ are the general normal tasks, but even they demand too large a processor load for computations. For the latest standard [9], a higher sample rate is needed to satisfy the modern requirements for energy quality, and this strongly contrasts with minimizing the processor load. A minimized processor load is a basic requirement for the developed flicker estimation method here, because several energy quality parameters also have to be estimated (RMS, THD, voltage asymmetry, etc.) at the same time; thus, the most efficient method has to be applied for flicker computation.

It appears that the coherent transformation steps of flicker estimation (e.g. cascade of filtering processes) can be handled together ensuring the increased efficiency and minimized processor load. According to the final strategy, the flicker computation is divided into two phases:

- In the first step, the raw samples are buffered into a temporary, 1-min buffer in order to be able to transform and generate the instantaneous signal (P_{inst}), which is a normalized signal and ready to direct flicker estimation through statistical process. If the 1-min buffer is full then *Thread 2* (cf. Fig. 8) starts the pre-processing method on raw samples and the generation of P_{inst} signal. The 1-min P_{inst} signal is transmitted to a 10-min FIFO memory for temporary buffering of the transformed (P_{inst}) data samples.
- In the second step, when the observation period lasts a longer time, e.g. for 1, 2, 5 or 10 min, and the P_{inst} FIFO has enough data samples, *Thread 2* automatically starts with the statistical method for flicker estimation, which does not load the processor as heavily as it does during the pre-processing.

In practice, this novel method works very efficiently for applying minimal memory and processor load—tests confirm the correct operation—as opposed to the

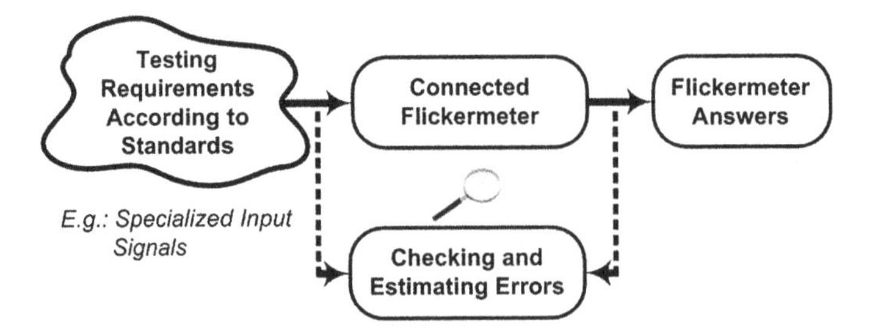

Fig. 9 Fundamental principles of the objective calibration method for testing flickermeters

brute-force methods operating without pre-processing with separated FIFO memory. (This one-phase method does not apply pre-processing during flicker computations, thus, it requires higher processor load and memory usage for data processing comparing to the new two-phase method; in addition, the maximum sampling rate is between 5 and 10 kHz according to the hardware and other running tasks.)

This new method facilitates a larger sample rate generation; thus, it is a chance to estimate flicker around 40 kHz on 6–12 channels. The large sampling rate is very important for the latest standard requirements for *Class* F_1 flickermeters, which use a larger sample rate then the *Class* F_3.

4.5 Installation of the Online Flickermeter, Serving as a Module to a Modern Network Diagnostics System

The newly developed online flickermeter module was installed into a network diagnostics system at the *Department of Electrical and Electronic Engineering at the University of Miskolc, Hungary*. This diagnostics system is used modern multi-core technology and multitasking task operations [19, 20]. The developed complex system to co-operate with the present online flickermeter module satisfies the standard [22] and is able to measure every required electrical parameter online and store data for offline analysis. The online module provides a separate thread for flicker analysis but this thread does not put impact on the other active threads of the diagnostics system.

Finally, a flexible diagnostics system as a whole was developed which is able to take complex measurements within the electrical network with the help of the necessary accessories, such as PC-based data acquisition system (card) and current and voltage transducers (cf. Fig. 13).

5 Flickermeters Testing by New Calibration System

This new calibration system was developed to test any type of flicker-meter and to confirm static and dynamic functionality. Standards declare the test specification of flickermeters and their responses to different network disturbances. The most important tests are the rectangular voltage changes and performance tests, the total response test with sinusoidal and rectangular test signals, the combined frequency and voltage changes test, the phase jump test, the distorted voltage test, the bandwidth test and the duty cycle test. The calibration method is based on the signal generation process in the tested flickermeter input comparing the output to the specification. This objective calibration method can handle not only this new flickermeter module but also any connected flickermeter when compared to the specifications (Fig. 9).

New simulation software generates complex modulation patterns [9] continuously for testing the module responses, comparing the specifications and checking the error of flickermeters. From the simulations, optimal sampling rates were determined for the specified flicker Classes F_1 and F_3 (F_2 is a laboratory certification category): for Class F_3 the optimal rate is 500 Hz; for Class F_1 the optimal rate is 12.5 kHz. Figure 10 gives simulation results at different sample rates, and the single most important criteria, the results of the phase jump test, are shown in Fig. 11. The 500 Hz (*Class* F_3) and 12.5 kHz (*Class* F_1) are the minimum limits of sample rates performing the standard [9] requirements, thus, the errors of measured flicker severities are under

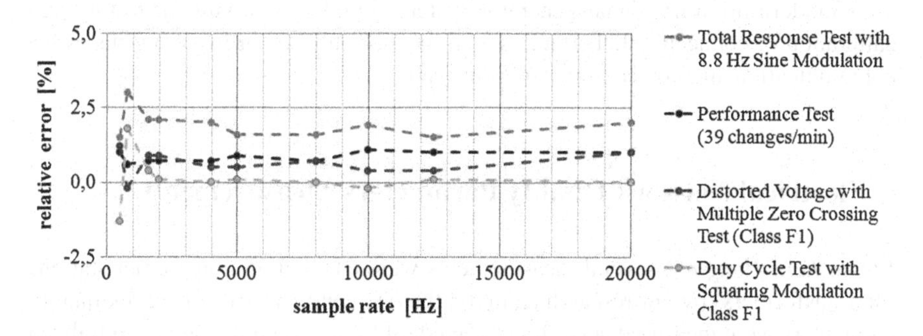

Fig. 10 Results of the measurements testing the new flickermeter at different sample rates

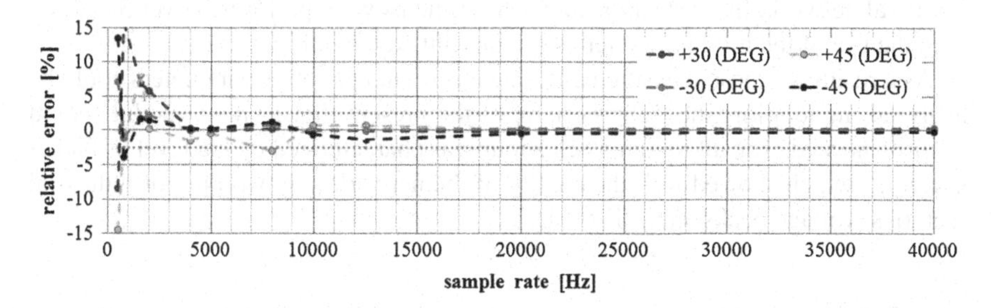

Fig. 11 Results of the phase jump tests for the new flickermeter module (*Class* F_1)

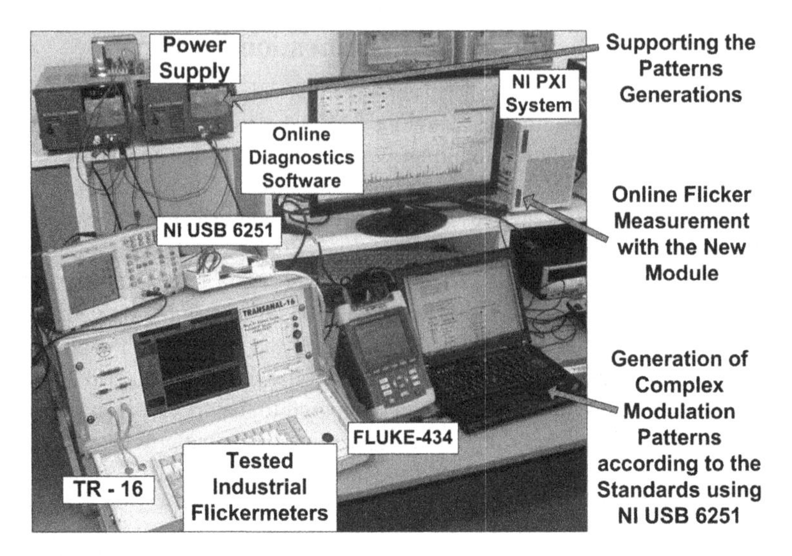

Fig. 12 Calibration setup for testing (*Class* F_3) flickermeters

3 % (According to the requirements, the errors of the measured flicker severities must be less than 5 %.) Confirming all simulation results, this new online flickermeter module operates with these sampling rates the best.

A modern arbitrary signal generator system (Fig. 12) was built for testing and calibration of connected flickermeters, defining specialized hardware interfaces for communication and power connections as well.

6 Electrical Network Quality Parameters Measurements

In *Northern Hungary*, several measurements were taken at 35 transformer stations (46 transformers) by our research team; on the one hand, electrical network quality parameters were measured according to standard [22], and on the other, centralized ripple control analysis were carried out for optimizing heat-storage power on the electrical network. In most cases, the measurements were performed over 3–7 days, with the use of the modular system structure illustrated in Fig. 13.

As a results of these measurements, the electrical network's parameters could be analyzed, in addition, the right working of the present flicker-meter module could be also checked. The all region's behaviour was characterized by the level of flicker severities which showed well the quality of the network (e.g. working of industrial factories, voltage peaks, etc.) (Fig. 14).

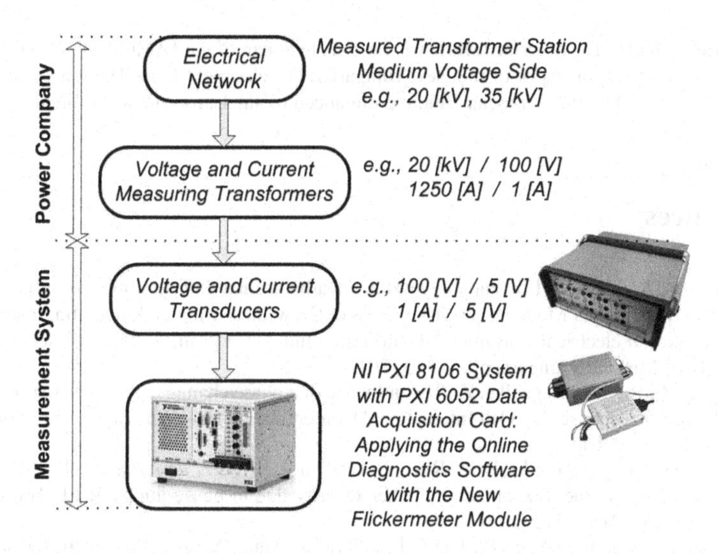

Fig. 13 Modular structure of the measurement system

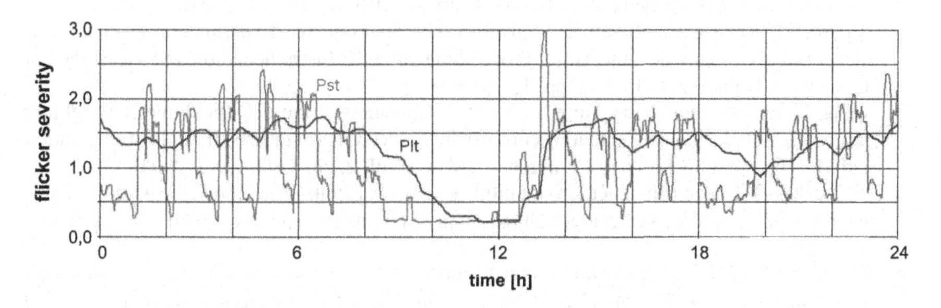

Fig. 14 Analysis results from a strongly disturbed electrical network by industrial factories

7 Conclusions

Diagnostic systems are vital in the measurement and evaluation of the electric power supply, especially for small and medium-sized enterprises and different transformer stations of power companies. Flicker is an important issue in power supply, and therefore a complex diagnostic system was developed containing a modern online flickermeter module. A flickermeter was developed that uses advanced multithreading technologies and whose performance meets the latest standard requirements. Its universal online module is modular, flexible and reliable, e.g. the sampling rate and its modular architecture can be changed. Furthermore, a new, objective calibration and testing method and system have been introduced, based on a newly developed modern arbitrary signal generator. Using this calibration, any type of flickermeter can be tested according to the objective standard regulations.

Acknowledgments The described work was carried out as part of the TÁMOP-4.2.2/B-10/1-2010-0008 project in the framework of the New Hungarian Development Plan. The realization of this project is supported by the European Union, co-financed by the European Social Fund.

References

1. Unhauzer, A.: Villamos Hálózati Fogyasztók Zavarhatásainak és Teljesítményprofiljainak Vizsgálata Új Mérési és Modellezési Módszerekkel (New measuring and modelling methods for diagnostics of electrical consumers' disturbances and power profiles). Ph.D. dissertation, University of Miskolc, Hungary, pp. 1–185 (2013)
2. Rogoz, M., Bien A., Hanzelka, Z.: The influence of a phase change in the measured voltage on flickermeter response. In: 11th International Conference on Harmonics and Quality of Power, pp. 333–337 (2004)
3. Gutierrez, J.J., Leturiondo, L.A., Ruiz, J., Lazkano, A., Saiz, P., Azkarate, I.: Effect of the sampling rate on the assessment of flicker severity due to phase jumps. IEEE Trans. Power Delivery **26**, 2215–2222 (2011)
4. Halpin, M., Cai, R., Jaeger, E., Papic, I., Perera, S., Yang, X.: A review of flicker objectives related to complaints measurements, and analysis techniques. In: 20th International Conference and Exhibition on Electricity Distribution—Part 1 (CIRED), pp. 1–4 (2009)
5. Peretto, L., Riva, C.E., Rovati, L., Salvatori, G., Tinarelli, R.: Experimental evaluation of flicker effects on human subjects. In: Proceedings of IEEE Instrumentation and Measurement Technology Conference (IMTC), pp. 1–5 (2007)
6. Wang, C.-S., Devaney, M.J., Yang S.-W.: Decomposition of the IEC flickermeter weighting curves (light flicker due to fluctuating current). In: Proceedings of the 21st IEEE Instrumentation and Measurement Technology Conference (IMTC), vol. 2, pp. 1378–1382 (2004)
7. IEC 61000–3-3: Electromagnetic compatibility—Part 3: Limits—Section 3: Limitation of voltage fluctuations and flicker in low-voltage supply systems for equipment with rated current ≤ 16A (2001)
8. IEC 61000–4-7: Electromagnetic compatibility (EMC)—Part 4: Testing and measurement techniques—Section 7: General guide on harmonics and inter-harmonics measurements and instrumentation, for power supply systems and equipment connected thereto (2002)
9. IEC 61000–4-15: Electromagnetic compatibility (EMC)—Part 4: Testing and measurement techniques—Section 15: Flickermeter—Functional and design specifications (2010)
10. Gallo, D., Langella, R., Testa, A.: Toward a new flickermeter based on voltage spectral analysis. In: Proceedings of the 2002 IEEE International Symposium on Industrial Electronics (ISIE), vol. 2, pp. 573–578 (2002)
11. Yang, X.X., Kratz, M.: Power system flicker analysis by RMS voltage values and numeric flicker meter emulation. IEEE Trans. Power Delivery **24**, 1310–1318 (2009)
12. Bucci, G., Fiorucci, E., Gallo, D., Landi, C.: Comparison among traditional and new digital instruments for the measurement of the light flicker effect. In: Proceedings of the 20th IEEE Instrumentation and Measurement Technology Conference (IMTC), vol. 1, pp. 484–489 (2003)
13. Ming-Tang, C., Sakis Meliopoulos, A.P.: Wavelet-based algorithm for voltage flicker analysis. In: IEEE 9th International Conference on Harmonics and Quality of Power, vol. 2, pp. 732–738 (2000)
14. Piekarz, M., Szlosek, M., Hanzelka, Z., Bien, A., Stankiewicz, A., Hartman, M.: Comparative tests of flickermeters. In: 10th International Conference on Harmonics and Quality of Power, vol. 1, pp. 220–227 (2002)
15. Gallo, D., Landi, C., Pasquino, N.: Design and calibration of an objective flickermeter. IEEE Trans. Instrum. Meas. **55**, 2118–2125 (2006)
16. Kang, W., Li, H., Yan, X., Zhang, L., Sun, F.: A modified demodulation method for flicker measurement. In: International Symposium on Power Electronics, Electrical Drives, Automation and Motion (SPEEDAM), pp. 765–768 (2006)

17. Ruiz, J., Lazkano, A., Aramendi, E., Leturiondo, L.A.: Analysis of sensitivity to the main parameters involved in the digital implementation of the UIE flickermeter. In: 10th Mediterranean Electrotechnical Conference (MELECON), vol. 2, pp. 823–826 (2000)
18. Clarkson, P., Wright, P.S.: Sensitivity analysis of flickermeter implementations to waveforms for testing to the requirements of IEC 61000–4-15. IET Sci. Meas. Technol. **4**, 125–135 (2010)
19. Bátorfi, R., Váradiné, Sz.A.: Electric network quality test with high accuracy by measuring instruments based on a new synchronizing procedure. Electrotechnics (Elektrotechnika) **105**, 5–9 (2012)
20. Bátorfi, R.: Electrical power quality and efficiency diagnostic system. In: Proceedings XIX IMEKO World Congress, Lisbon (2009)
21. Bollen, M.H.J., Gu, I.Y.H.: Signal Processing of Power Quality Disturbances. Wiley-Interscience, Piscataway (2006)
22. EN 50160: Voltage characteristics of electricity supplied by public electricity networks (2010)

Presenting a Logistics Oriented Research Project in the Field of E-marketplace Integrated Virtual Enterprises

Róbert Skapinyecz and Béla Illés

Abstract The aim of the chapter is to present an ongoing research project in the field of virtual (or extended) enterprises, focusing mainly on perspectives of logistics (especially freight transport) and e-commerce. The novelty of the research lies in the introduction of certain quality-management methods into the field, which so far have not been implemented for electronic-commerce based logistics services. In addition to the presentation of the above mentioned concepts, the chapter also gives a practical overview of the utilization of e-marketplaces in the logistics industry, supplemented by certain practical examples.

1 Introduction

The significance of cooperative logistics systems has continuously been on the rise in recent years, mainly due to an increased need for cooperation in logistics on the inter-organizational level and also to the continuous advance of the related IT infrastructure. One important class of such cooperative systems is represented by what are called 'virtual' (or 'extended') enterprises.

A virtual enterprise itself is generally defined as a group of geographically dispersed, self-autonomous organizations and enterprises which temporarily operate in a cooperative manner in order to achieve a mutual goal or a set of goals while the individual organizations keep their complete autonomy. Naturally, this definition only gives a conceptual frame, as in practice several different types of organizations can fulfil the above mentioned requirement (e.g. business clusters, supplier networks, etc.). However, a main characteristic of true virtual enterprises is that the individual SMEs (small and medium-sized enterprises) inside the organization have to be

R. Skapinyecz (✉) · B. Illés
Department of Materials Handling and Logistics, University of Miskolc,
Miskolc-Egyetemváros 3515, Hungary
e-mail: altskapi@uni-miskolc.hu

G. Bognár and T. Tóth (eds.), *Applied Information Science, Engineering and Technology*, Topics in Intelligent Engineering and Informatics 7, DOI: 10.1007/978-3-319-01919-2_11,

able to completely preserve their own autonomy, which in other words means that a high level of industrial and economic cooperation has to be realized almost entirely through means of information sharing. However, it is a significant challenge in such systems to facilitate the required level of trust among the different participants, which is a basic requirement for the proper operation of any cooperative system.

In regard to this problem, the chapter introduces an emerging concept for creating a risk-evaluation method that can be applied in electronic marketplace based virtual enterprises, mainly concentrating on the fields of logistics and transportation. Such a tool could be a great asset for both the customers and the potential participants of a virtual logistics organization, as it could provide an objective view regarding the reliability of the overall system. In addition, in its first sections the chapter also gives a useful summary of some of the basic aspects of e-marketplaces used in the field of transportation.

2 Electronic Marketplaces in the Freight Transport Industry

The widespread adoption of electronic marketplaces has taken place mainly during the last two decades, parallel with the large scale expansion of the Internet (it has to be noted that certain types of electronic commerce had already been active on closed networks before). The main aim of these marketplaces is that of the distribution of the manufacturers' production capacities among the buyers, according to the laws of the free market (hence the designation). This results in numerous benefits and some drawbacks as well. However, the widely accepted view is that they are mainly beneficial for small and medium sized enterprises. In the following, the main characteristics of these e-marketplaces are presented, together with some practical examples from the transportation industry.

2.1 General Introduction of Transportation Marketplaces

Electronic marketplaces can be viewed as scenes of the e-commerce realized between the vendors and the customers [1]. The literature makes a distinction between B2C (Business-to-Customer) and B2B (Business-to-Business) e-marketplaces. The individual customer is usually familiar with the first of the two types (through the use of electronic stores like Amazon.com, on-line auction websites like E-bay, electronic stores specialized in the sale of specific products, etc.). However, this chapter mainly focuses on the second type.

From the aspect of the maintainers of B2B e-marketplaces, we can distinguish between different models: the buyer-oriented model, where the maintainer of the e-marketplace is an individual company or consortium with significant customer potential, who maintains the marketplace for its possible suppliers; the seller-oriented model, which is maintained in the previous way but for the potential customers of the company or consortium; or the mediator model, where the marketplace is maintained by an independent third party [1]. The e-marketplaces created for the sale of transportation services can be categorized into the 1st and 3rd groups. Based on their working mechanisms, we can further categorize these into the following groups [2]:

- clearing houses,
- freight exchanges,
- auction houses.

In a clearing house, the customers (shippers) post their needs, while the vendors (carriers) post their free capacities. If an actor finds a suitable solution on the other side, then negotiation can start between the two parties, with further contact taking place outside of the clearing house. This process is based on a database that consists of either the loads posted by the shippers, or the transport capacities posted by the carriers [2].

Freight exchanges are in many ways similar to clearing houses, but in this case, the negotiation usually takes place on the e-marketplace. Therefore, such institutions usually provide a range of additional services that can help in the efficient handling of the negotiation process. The latter usually involves the use of specialized agents on both sides and these can also be software tools with artificial intelligence, depending on the level of automation which is implemented on the marketplace [2].

For the realization of long-term transportation contracts, the most widely used method is the use of combinatorial auctions. In such negotiations, the vendors offer not only a single item, but a set of different items for the potential customers. Then the customers can make their bids for optional subsets of the initial set, which has the consequence that a combinatorial auction can have multiple winners. However, it is a requirement that all of the items have to be sold during the auction, moreover that a single item can only fall into the possession of one bidder.

In the auction houses which serve the realization of long-term transportation contracts, the auction proceeds in the opposite way: a customer (shipper) offers the routes on which they require transportation services. Then the vendors (carriers) post their bids for the routes which they would like to serve. In this case, the bids consist of the payments the vendors would like to receive in return for the service provided (e.g. the transportation of the customer's goods), therefore it is natural that the combination with the lowest total price will be the winner (at least in the majority of cases).

The importance of the auction's combinatorial nature is that from the aspect of a given carrier, the routes offered can frequently be related to each other [3]. A typical example can be the case where a carrier can insert several of the routes into its already existing network in a way which not only generates more profits, but also decreases the number of empty vehicle trips in the network. A special variant of the previous case is when the chosen routes can be inserted into a round trip; in such cases, the carrier cannot only increase its income, but can also decrease the specific transportation costs (due to the lower number of empty vehicle trips within its network). It follows that the carriers can seta lower pricing level if they can choose

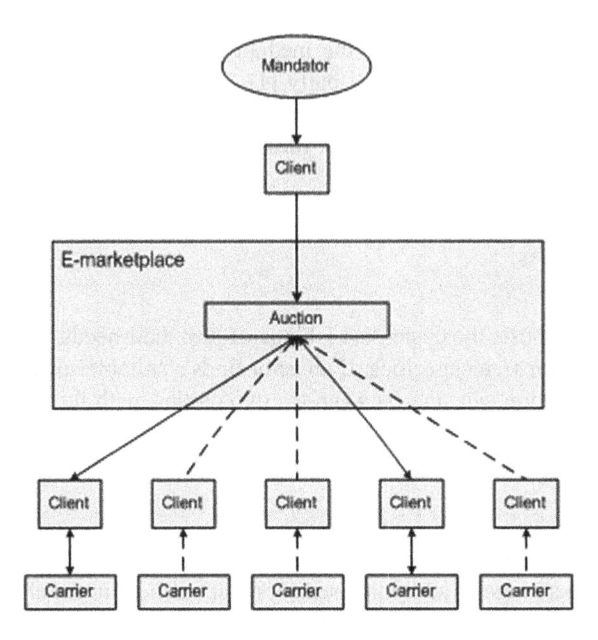

Fig. 1 General implementation of the auction-based e-commerce inside the freight transport industry

the most appropriate combination of the sets of the routes which is posted by the shipper. Thus it is evident that the use of the auction mechanisms described above can be highly beneficial for both sides (Fig 1).

In practice, combinatorial auctions have been in use in the freight transport industry for almost two decades. Nevertheless, it has to be noted that the method is primarily used in conjunction with the realization of long-term transportation contracts, as short-term assignments usually contain only a single route, and therefore the use of combinatorial auctions would not make any difference.

2.2 Examples in the Industry

One of the first users of combinatorial auctions in the freight transport industry was Sears Logistics Services (SLS), an American company that was able to save more than 84 million dollars in the early 1990s by applying the method for choosing its logistics service providers for its network of 854 routes [4]. This example was later followed by other companies, like Home Depot (also American), the world's largest home improvement retailer, which used the method with prominent success at the beginning of the new millennium [4].

In the above examples, the marketplace was operated by a single customer; therefore these can clearly be sorted into the first group of B2B e-marketplaces, which represents the buyer-oriented model. Later, independently operated e-marketplaces also came into existence, where numerous customers (shippers) and vendors (carriers) are present at the same time. Such marketplaces are operated by entities like Freightmatrix, Freight-Traders, Transcore, Transplace and numerous other enterprises not listed here [2]. In many cases, these marketplaces provide other value-added services as well, e.g. the use of optimization software for the carriers, or the ability of cargo tracking and the use of decision support tools for the shippers. Moreover, in the previous examples the marketplace usually supports the realization of both short-term and long-term contracts [2]. In addition, there are virtual institutions which expressly concentrate on the realization of short-term assignments. A good example for the latter is the Turkish ESO Logistics Center, which was founded in 2003 near the city of Eskisehir [5]. This is a physically existing logistics center, one which maintains its own electronic marketplace for the purpose of supporting the realization of the freight assignments that are implemented with the help of its services. ESO also uses an auction mechanism, but as the assignments in this case are single items, so the auction is not a combinatorial one [5].

All of the above listed organizations almost exclusively concentrate on road-based transportation, but examples can be found for the uses of e-marketplaces in other branches of transportation as well. Such examples are GoCargo.com in maritime transport, or Global Freight Exchange (lately Descartes GF-X Exchange) in the air freight transport [2]. The latter industry is also characterized by the formation of the so-called Cargo Community Systems (CCS), which mainly concentrate on supporting small and medium-sized freight forwarders with different IT solutions, based on favourable and cost-effective conditions (e-marketplace, EDI, etc.).

3 Introduction of E-marketplace Integrated Virtual Enterprises

Virtual enterprises are receiving significant attention in modern logistics research [6]. The primary reason behind this tendency is that the dominant part of today's logistics systems operates in a network-like manner, which is usually based on some form of cooperation among self-autonomous organizations. The logical consequence of this practice is that the more highly organized logistics networks show similarities in their operation to virtual enterprises. However, a constantly recurring problem of the field is that a universally accepted definition of the virtual enterprise is still missing in the literature, while a significant number of individual approaches to the problem can be found. From a certain point of view, this can be considered natural, as all existing organizations have their unique characteristics; however, from the aspect of the ongoing research activities, it would be beneficial to lay down the definitive characteristics of virtual enterprises.

The following section of the chapter attempts to provide the basis for the above mentioned goal, starting out from the most general definition of virtual enterprises.

Then it introduces the concept of the e-marketplace integrated virtual enterprise, which may be one answer for the problem of how to implement a virtual organization in practice (it has to be mentioned that, according to the best knowledge of the authors, although the term 'e-marketplace integrated virtual enterprise' has not been introduced before now, several organizations operate in a similar manner).

3.1 General Definition of Virtual Enterprises

According to the most widely accepted general definition, the virtual enterprise can be considered as a self-organized group of individual enterprises (tipically SMEs), often covering a large geographical area and working in many ways as a single organization, trying to achieve a common set of goals for the participating enterprises. Of course, this general definition can be valid for a large variety of different network-like organizations. However, according to the views of the author, one main aspect for the categorization of the different network-like organizations could be the fulfilment of the criteria that the individual enterprises inside the organization must have their own autonomy preserved. Therefore, the cooperation among the individual units has to be realized through the coordination of the resources and information, rather than the use of a single controlling entity.

The latter requirement could be fulfilled through the implementation of some form of a distributed intelligence based network where the role of the third party is "just" to maintain the information system which connects the members of the virtual organization. The coordination may be realized through the use of the common infrastructure, and through the implementation of a commonly accepted set of rules which partially govern the behaviour of the individual members inside the organization. As can be seen, the e-marketplace integrated virtual enterprises (which can also be derived from the model of the electronic store) fit well into the previous concept.

3.2 Concept of the E-marketplace Integrated Virtual Enterprise

The essence of the e-marketplace integrated virtual enterprise—as the definition clearly shows—is that the distribution of the organization's resources takes place on an electronic marketplace. In the simplest case, this marketplace may be a single web page which creates the possibility for small and medium-sized enterprises to find their optimal partners (usually suppliers or service providers) through this medium.

Naturally, in this form the system would not be significantly more than a closed electronic marketplace maintained for a certain group of enterprises. However, it was clearly demonstrated in the second section that such systems can also effectively support the accomplishment of more complex inter-organizational tasks, mainly by providing different complementary services. In the field of logistics, such services may be route optimization, cargo tracking and tracing, allocation of storage capacities, implementation of chapterless data communication between the parties (most often through the use of EDI), etc. With the integration of such services, the e-marketplace may truly become capable of the organization of complex inter-organizational operations, especially if the negotiation processes are highly automated (generally through the use of intelligent agents).

Besides the previous possibilities, commonly accepted bidding and transaction rules may be introduced on the internal marketplace, which alone can realize a form of coordination in an implicit manner. However, this solution also leaves the decision autonomy of the individual parties intact, and the internal marketplace can be viewed as a medium that creates the cohesion among the participating SMEs. As can be seen, in such a model every decision is made in the scope of the individual members, while from an outside point of view, the whole system truly resembles a single organization. In other words, such systems can be categorized into the family of holonic organizations.

The following section will demonstrate a system operating in this manner in the freight transport industry.

3.3 Concept of the E-marketplace Integrated Virtual Transport Enterprise

The e-marketplace integrated virtual transport enterprise can be defined as a special case of the previously outlined enterprise model. The main characteristic of such an organization is that the system mainly concentrates on the optimal distribution of transport capacities (complemented by other auxiliary services). As could be seen in the second section, electronic marketplaces themselves have a fairly long history in the industry. However, these alone cannot be considered to be virtual enterprise, for they lack a commonly accepted framework which could serve as a basis of higher level cooperation. Therefore, in order to create a real virtual transport enterprise, first this framework has to be outlined and implemented, according to the willingness of the participants. The framework, in its most basic form, should contain the following elements [7]:

- qualification of the transport companies (carriers) according to their real performance and the utilization of these results in determining the group of the participating carriers,
- utilization of a commonly accepted auction mechanism (ideally a closed, single-round combinatorial auction),
- use of a single and common optimization tool—provided by the maintainer of the marketplace—that helps in the formulation of all the bids,
- use of a common client program for accessing the system, also provided by the maintainer, which, among other services, realizes a complete EDI connection among the members of the virtual enterprise.

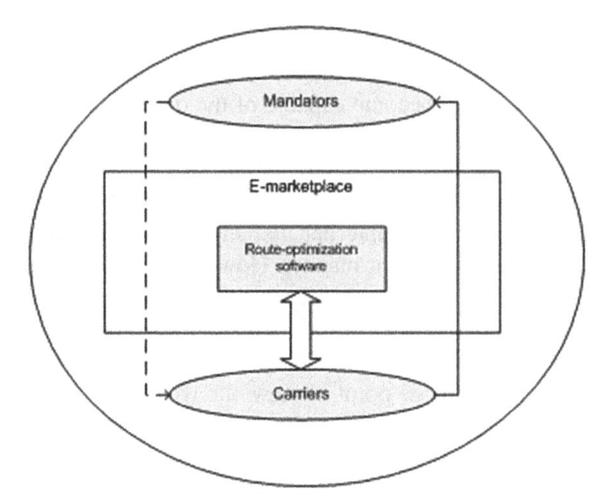

Fig. 2 Working concept of the e-marketplace integrated virtual transport enterprise

The framework outlined above formulates the minimal requirements, but naturally, other rules and coordinative tools may be integrated into the system as well. However, all improvements of the basic system should be introduced in a common manner in order to provide equal conditions to all members. The basic functionality of the outlined system is presented in Fig. 2, where the dotted line represents the call for bids by the mandators (who themselves are members of the virtual organization), while the continuous line represents the bidding processes.

One practical advantage of the model presented is that the real system can be introduced through a sequence of consecutive steps, where the individual members may jointly determine the level of integration inside the organization, taking into account previous experiences. This is important because in many cases, the greatest problem arises in the practical implementation of the given virtual enterprise model, as the future members of the organization are often unable to see the concrete benefits of their participation, while they might also have concerns about the effective functioning of the system. However, the proposed flexible architecture allows a gradual implementation of the system, starting from the introduction of a freely accessible and simple electronic marketplace, which does not require the introduction of any restriction among the members.

Another advantage of the concept is that it realizes a two-level optimization approach: on the first level, the carriers optimize their routes during the bidding process, while on the second level, the mandators choose the optimal offer from the incoming bids. This approach inherently realizes distributive computing, while in the meantime it improves the overall efficiency of the utilization of the given transportation capacities.

4 Possibilities of Optimization

4.1 Possible Approaches

Regarding the possibilities of optimization in virtual organizations, we can mainly distinguish between centralized and decentralized optimization methods [8]. The centralized methods, as their name suggests, are based on the centralized distribution of the organization's resources. In some ways, the combinatorial auction mechanism can also be categorized into this type of method, as in this case, the customer chooses the optimal offer from the bids of the suppliers or service providers. This type of approach has many advantages but also has some weaknesses. Most notably, in such mechanisms it is difficult to achieve cooperation among the different suppliers and service providers [8].

The decentralized optimization methods seem to be more suitable for implementation in a holonic organizational environment, as they are naturally based upon distributive computing architectures. In such methods, the solution usually emerges as the result of a complex negotiation process among different autonomous agents. Many theoretical and practical models follow this approach, but their complexity has prevented their widespread adoption in the industry so far.

In the following section, a somewhat different method will be introduced, which is designed to integrate the centralized and decentralized approaches into a single mechanism (a special type of combinatorial auction). In this way, the work distribution among the carriers may be significantly increased, while the main process can still be easily handled.

4.2 A Proposed Optimization Method

The main characteristic of the proposed method is that it allows for the routes to be served by multiple carriers. This is achieved through handling the transportation capacities offered as dynamic variables in the optimization process, which can be modified within certain limit values defined by the given carrier. However, it is important to note that the combinatorial bids are only valid in their complete form, therefore the lower limit of the capacity offered has to be accepted on each route which the carrier has applied for, otherwise the complete bid will be rejected.

The point of the process outlined above is that it combines the resources of the bidding carriers according to the needs of the shipper. However, the willingness for cooperation on the carriers' side is an important factor in the process, as they define the variability of the capacities offered. Therefore, the effectiveness of the system will mainly depend on the flexibility of the carriers.

For the shippers, the primary advantage of the method presented is that if the performance of one of the assigned carriers becomes temporarily reduced, another carrier that is present on the affected route(s) would be able to easily compensate for the missing performance. Furthermore, this approach would also ensure that the shipper would not get into a position where it would overly depend on a limited number of carriers. From the aspect of the carriers, the method would mainly benefit the small and medium-sized service providers, since through this approach they may be able to enter into routes which would be otherwise unattainable for them.

In addition to the previous advantages, the system may also contribute to the more efficient utilization of the available resources of the participating carriers, which eventually culminates in better energy management for the whole system. This aspect of optimization has outstanding significance today, when increasing energy and fuel prices represent a major barrier for developing economies, especially for countries and regions which depend heavily on foreign energy sources. Thus, the realization of the proposed model not only increases economic efficiency, but also reduces energy dependency and the environmental load of the freight forwarding industry.

As was mentioned before, in order to keep the trust of the customers, it may be necessary to bind the vendor's entry into the market to certain quality requirements. Moreover, modifying factors can be deduced from the quality performance of the vendors, which can be applied for a complex evaluation of the incoming bids. Such factors include the development level of the carrier's information system *(i)*, the reliability that characterizes the carrier *(r)* and the punctuality that characterizes the carrier's operations *(p)*. The values of these factors can be determined according to a predefined scale. Naturally, these are just a few examples and numerous other factors can be introduced for the purpose of evaluating the bids of the carriers.

It may also be necessary to formulate certain requirements for the shippers as well, such as the precise disclosure of information related to their assignments. This can be important from the aspect of the carriers, as they can only utilize their resources effectively if they are aware of the exact transportation demands on the individual routes. These demands can be expressed with the amount of material flow work that is expected by the shipper on the individual routes (given e.g. in tons/day, tons/week, tons/month, or in other similar formats).

Naturally, numerous other parameters have to be defined in connection with the transportation demand on a given route (such as types of the particular unit load carriers, types of the particular vehicles, types of the transported goods, etc.), but these questions are not investigated in the current chapter. Instead, the chaper continues with the mathematical composition of the presented method.

The first step is the formulation of the auction call by the customer (shipper), which may be in the following format:

$$S = \{l_{xj}, l_{yj}, l'_{xj}, l'_{yj}, q_{szj}, t\},\ j = 1, 2, \ldots, m \tag{1}$$

where:

l_{xj}, l_{yj} the geographical coordinates of the starting location of the *j*th route,
l'_{xj}, l'_{yj} the geographical coordinates of the ending location of the *j*th route,
q_{szj} the nascent transportation capacity requirement on the *j*th route (given for example in tons/week),
t the duration of the assignment.

The carriers send their bids in response to the auction call. Ideally, each bid is constructed with the help of the jointly used route-optimization software, while all the necessary data are provided by the customer. The bid structure of the carriers is the following:

$$F_i = \{[q_{ij}, q_{Minij}, q_{ij}^F, c_{ij}^F], b_i\}, \; j = 1, 2, ..., m \tag{2}$$

where:

F_i the bid of the ith carrier, $i = 1, 2, \ldots, n$,

q_{ij} total amount of the transportation capacity which the ith carrier offers for the jth route (zero value is allowed),

q_{Minij} amount of the transportation capacity offered by the ith carrier for the jth route which the customer has to accept in order to keep the bid of the ith carrier,

q_{ij}^F standard unit of transportation capacity that serves as the basis of freightage calculation at the ith carrier and for the jth route,

c_{ij}^F the ith carrier's specific freightage, defined for the jth route and for the given time interval, i.e. the cost of a single unit of transportation capacity,

b_i logical variable that defines the actual status of the ith carrier's bid during the process of optimization (it equals to 1 if the bid is part of the solution, and it becomes zero if the bid is rejected).

Now we can define the objective function that has to be optimized and which minimizes the costs of the customer ($q_{ij} - \Delta\, q_{ij}$ represents the portion of a certain bid that is accepted by the customer on a given route):

$$C = \sum_{i=1}^{n}\sum_{j=1}^{m} c_{ij}^F [\frac{q_{ij} - \Delta q_{ij}}{q_{ij}^F}] t b_i = \min \tag{3}$$

The boundary conditions ensure the realization of the mechanism of resource allocation described above:

$$q_{Min_i j} \leq q_{ij} - \Delta q_{ij} \leq q_{ij} | \forall i, j \tag{4}$$

$$\sum_{i=1}^{n} (q_{ij} - \Delta q_{ij}) b_i = q_{SZj} | \forall j. \tag{5}$$

5 Introduction of Quality Parameters

It is clear from the above presented model that one of the most fundamental tasks during the optimization processes is that of the objective evaluation of the participating SMEs from a risk-management perspective. In order to do this, fundamental evaluation metrics have to be introduced. Basic parameters of such a metric for a transportation enterprise may be:

- punctuality, which describes the deviation of the actual delivery dates from the planned values forby a certain carrier or freight-forwarder (denoted by 'p'),
- reliability, which describes the general failure rate in the delivered goods for a certain carrier (denoted by 'r'),
- calculability, which describes the deviation in the amount of the delivered goods from the planned values for a certain carrier (denoted by 'c').

The objective function below serves as an example of how to incorporate such a metric into the optimization process (in this function, each risk parameter can increase or decrease the total cost of the given bid of the ith service provider):

$$C = \sum_{i=1}^{n} \sum_{j=1}^{m} c_i r_i p_i c_{ij}^{F} [\frac{q_{ij} - \Delta q_{ij}}{q_{ij}^{F}}] t b_i = \min \tag{6}$$

The greatest obstacle in the quantification of the proposed risk parameters is that in many cases the acquisition of the necessary data for the long-term evaluation of the logistics service providers is an almost impossible task. This is the primary reason behind the proposed application of the so called 'process-capability' model, which has been in use in various industrial fields for many decades by now, and has proven itself many times under real-life conditions [9]. In short, the essence of the approach is that the deviation of many key parameters in an industrial process can be adequately described with the use of a few typical statistical distributions (most often with the normal distribution). However, this also means that an adequate sample of the examined parameter can be used for making assumptions about the characteristic deviation (σ) and the mean value (μ) for this parameter. The concept of process capability uses a method where the obtained values of σ and μ are compared with the pre-defined tolerances of the process examined, in this way providing an objective assumption about the general 'capabilities' of the process. In practice, this comparison is made via the calculation of the process capability and critical process-capability parameters (c_p and c_{pk}, respectively) [9]:

$$c_p = \frac{T}{6\sigma} = \frac{OG - UG}{6\sigma} \tag{7}$$

$$c_{pk} = \min\{c_{po}; c_{pu}\} \tag{8}$$

In the first expression, *OG* is the upper tolerance value and *UG* is the lower tolerance value for the examined parameter, therefore the difference between the two gives the width of the allowable range for the parameter. In the second expression, c_{po} and c_{pu} are defined as (OG-μ) and (μ-UG) respectively (the "upper" and "lower" part of the allowable range), therefore c_{pk} will equal to the smaller, thereof more critical value. It is very important that this approach has started to expand into the

Table 1 Value of a given risk parameter in accordance with the associated σ levels

At 1.5-fold shift of the mean value μ

c_p	c_{pk}	Sigma level	Failure per million	Value of p
1.00	0.50	3	66810.6	1.0668
1.33	0.83	4	6209.7	1.0062
1.67	1.17	5	232.7	1.0002
2.00	1.50	6	3.4	1

service sector, and therefore into the field of logistics services as well, even though the normal distribution alone is rarely applicable in these areas (usually normalization is needed). In the manufacturing industries, these process capability parameters have already been in use for many decades, combined with various evaluation metrics. One of the most widely used of the metrics is what is called the 'six sigma' method, which is significant for two reasons: first, it is widely used in many different industries and a great number of companies even make its use mandatory for their suppliers—therefore it is already used by certain logistics-service providers as well. Secondly, in this metric discrete failure rates are connected to each sigma level which can serve as the starting point for calculating the additional cost of a less-dependable logistics service. For the actual cost calculation, the simplest and most straightforward approach is the linear connection, which is also used below to find the value of the risk parameter '*p*' (values for c_{pk} are calculated with assuming—due to long term distortions—a 1.5-fold shift in the mean value, which is a typical calculation method in the industry): It can be seen that this approach greatly encourages the service providers to increase the quality of their services, as the cost of their bids can be significantly affected by even a single difference in the sigma level (note the difference between Levels 3 and 4). Naturally, it has to be noted that the linear method can only be applied for the reliability parameter '*r*', where the cost of the failures is more or less linearly related to the average number of faulty products in the shipments. However, this does not affect the usefulness of the approach in general, as more sophisticated relation functions can be worked out for the other parameter types (Table 1).

The previously presented method can serve as an important decision making tool for the participants of a virtual organization; however, it does not provide a risk-oriented evaluation of the entire system. For this purpose, the author proposes the introduction of a general variable, which can be called a 'system capability parameter' (derived from the concept of 'process capability' for individual processes). As its name suggests, this parameter would deal with the entire organization in an implicit manner, viewing it as a single and complex process. Thereby it is obvious that this single parameter, serving as an evaluation tool of the entire virtual enterprise, would be the function of the individual risk parameters introduced into the system:

$$P = f(c_1 \ldots c_i \ldots c_n, r_1 \ldots r_i \ldots r_n, p_1 \ldots p_i \ldots p_n). \quad (9)$$

Table 2 Value of P in relation to the σ level of the overall system and to the σ levels of its components

Sigma level of components	Sigma level of the system	Failer per million for the system	Value of P
≥3	3	66810.6	≤1.0668
≥4	4	6209.7	≤1.0062
≥5	5	232.7	≤1.0002
≥6	6	3.4	1

The primary difficulty in this function is how to generalize this for all types of virtual logistics companies though the sizes and the types of services may vary significantly from organization to organization. One starting point in this research may be the fact that P can also be obtained by comparing the estimated cost of each mandate in the organization with its actual realized cost, which (in theory) can only be higher due to certain failures and inefficiencies during the realization:

$$P \sum_{k=1}^{p} C_k = \sum_{k=1}^{p} C_{Fk} \tag{10}$$

where:

- C_k is the estimated cost of the kth mandate in the examined period,
- C_{Fk} is the realized cost of the kth mandate in the examined period.

This latter method can be used both in operating real-life enterprises as a systematic control tool and in simulation models as a testing and research method. In the following stages of this research, the expectation is that by combining the use of the two approaches on a wide scale of simulation models, the first function can be determined for a generic virtual logistics enterprise with adequate precision, providing the possibility of making accurate predictions in real-life virtual logistics companies that apply the proposed risk-management concept. Table 2 clearly represents, the utilization of the P parameter could also serve as an implicit quality-parameter for a given virtual enterprise, as it is strongly related to the σ levels that are achieved at the individual components (member enterprises), while it also defines an overall σ level for the entire system. One aim of the research is to clarify this relationship, as it can have significant implications for the industry.

6 Summary

The chapter presented the basic concepts of an ongoing research project related to virtual logistics enterprises, as well as to cooperative networks and electronic marketplaces. The primary aim of this presentation was to introduce a possible new method for the evaluation of the performance of such entities, mainly from a quality perspective. Hopefully, this method may lead to the development of a general evaluation tool, one that could be used with significant success for the evaluation of virtual logistics organizations, at least from the aspects of risk and quality management. As the main difficulty of creating cooperative networks lies in achieving a significant level of trust and reliable information sharing among the different participants, such tools may play an important role in the future proliferation of network-based organizations.

Acknowledgments The described work was carried out as part of the TÁMOP-4.2.2/B-10/1-2010-0008 project in the framework of the New Hungarian Development Plan. The realization of this project is supported by the European Union, co-financed by the European Social Fund.

References

1. Kacsukné, B.L.: Modellek és megoldási algoritmusok logisztikával integrált elektronikus piacterekhez, PhD thesis, University of Miskolc, Department of Materials Handling and Logistics, http://www.hjphd.iit.uni-miskolc.hu/files/disszertaciok/Kacsukne_PhD_disszertacio.pdf (2005). Accessed: 18Jan 2010
2. Nandiraju, S., Regan, A.C.: Freight transportation electronic marketplaces: a survey of the industry and exploration of important research issues, Institute of Transportaion Studies, University of California, Irvine.http://www.its.uci.edu/its/publications/chapter/CLIFS/UCI-ITS-LI-WP-03-12.pdf (2003).Accessed 24 Jan 2010
3. Sheffi, Y.: Combinatorial Auctions in the Procurement of Transportation Services. Interfaces **34**, 245–252 (2004)
4. Elmaghraby, W., Keskinocak, P.: Combinatorial auctions in procurement. School of Industrial and Systems Engineering. Georgia Institute of Technology, Georgia (2002)
5. Ağrali, S., Tan, B., Karaesmen, F.: Modeling and analysis of an auction based logistics market. Eur. J. Oper. Res. **191**, 272–294 (2007)
6. Cselényi, J., Illés, B.: Anyagáramlási rendszerek tervezése és irányítása, Miskolci Egyetemi Kiadó, ISBN 963-661-672-8 (2006)
7. Skapinyecz, R., Illés, B.: Optimization of e-marketplace integrated virtual enterprises. Conference Proceedings, 4th International Doctoral Students Workshop on Logistics, University of Magdeburg, ISBN 978-3-940961-57-0 (2011)
8. Karageorgos, A., Mehandijev, N., Weichhart, G., Hämmerle, A.: Agent-based optimisation of logistics and production planning. Eng. Appl. Artif. Intell. **16**, 335–348 (2003)
9. Illés, B., Glistau, E., Machado, C.N.I.: Logistik und Qualitätsmanagement, 1. Auflage, Budai Nyomda, ISBN 978-963-87738-1-4 (2007)

Agribusiness Clusters and Logistic Processes Through the Example of Hungary

Balázs Illés and Béla Illés

Abstract This chapter arises from the interdisciplinary cooperation of researchers in regional development and logistic services. It is important to use the agricultural resources of a country in the most efficient way. Regional development is based on the proper coordination of resources, tools and methods. We present a basic summary of the logistic processes which should be considered when decision makers are to establish a hub-and-spoke type business network. As an example, we consider waste management in the agricultural and food industry. The objects of the study are [1] to analyze passive resources in Hungary, [2] choose an appropriate regional development theory and tools for the Hungarian economy, [3] implement the tools in agricultural and food industry waste management, and [4] represent the logistics in energy farm networks.

1 Motivation

Every research must have a personal motivation to drive it, and ours is the regional development in Hungary. Our home region of North Hungary is particularly underdeveloped in terms of GDP per capita and employment among the other EU-27 regions (9,700 Euros per capita, 40 % of GDP PPS per inhabitant in % of the EU27 average) (Source: Hungarian Central Statistical Office (HCSO). An appropriate model is needed to enhance regional development. The authors are enthusiastic about agricultural development because our culture is certain that the value of the soil is bound up with the value of "homeland". We decided to examine the structure of agriculture

B. Illés (✉)
Institute of Management Science, University of Miskolc-Egyetemváros, Miskolc 3515, Hungary
e-mail: szvilles@uni-miskolc.hu

B. Illés
Department of Materials Handling and Logistics, University of Miskolc-Egyetemváros, Miskolc 3515, Hungary

G. Bognár and T. Tóth (eds.), *Applied Information Science, Engineering and Technology*, Topics in Intelligent Engineering and Informatics 7, DOI: 10.1007/978-3-319-01919-2_12,

and its contribution to GDP and employment. The improper usage of soil has to do with the management of the individual farmers and the lack of useful methods and tools that can be adopted to the Hungarian situation.

2 A Glance at the Agricultural Situation in Hungary

2.1 Natural Capital, Labor and Machinery

The main problem is that agricultural land is not used as it should be. After collecting data about the situation of agribusiness in Hungary, we did an analysis to find out the reason for the lack of co-operation in this sector. The analyzed data were the GDP share of agriculture, employment in agriculture and the sector's structure. The analysis showed that the share of agricultural land within the country is high. When we consider the agro-forestry sector, it represents 78 % of the whole area of Hungary in Fig. 1.

The area used for only agricultural production is about 57 % in Hungary. This means that of the 27 countries of the European Union, Hungary, Greece and Denmark have the highest rate of agricultural land use. Later on in this chapter we will introduce that our first insight was mistaken when we emphasized the importance of this ratio. Considering the fact that these countries are not the biggest of the federation, it is also obvious that it is only a relatively high figure, because the larger nations have more land in total (Fig. 2). Our correlation analysis showed the patterns of agricultural GDP and the size of agricultural land, and obviously "the bigger the better" idea is working in this case.

Hungary is the 8th biggest country in Europe in terms of agricultural land. From our point of view the landside is one of the main factors that can contribute to the added value in this sector. The other factor is the labor force working in the field of agriculture. Hungary is ranked 9th in this comparison.

If we use an extremist model where the output of agriculture can be calculated by the land size and by the workforce we would expect the output of Hungary's agriculture will follow the aforementioned pattern and the position of the country

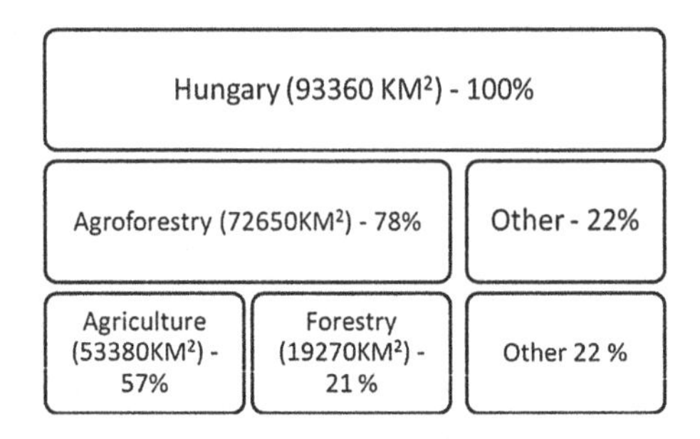

Fig. 1 Land use in Hungary (*Source* Hungarian central statistical office (HCSO)-Földhasználat művelési ágak és gazdaságcsoport szerint -Utilized land by the types of cultivation—05. 10. 2012)

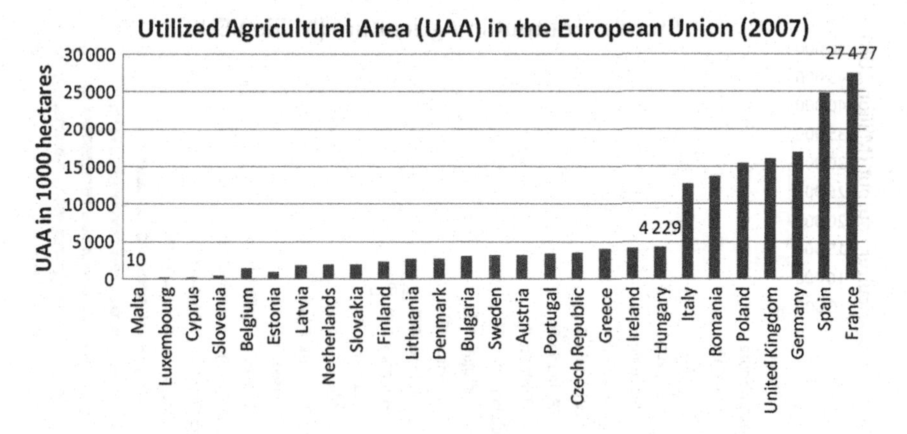

Fig. 2 (*Source* EUROSTAT—farm structure YB2013)

will be somewhere in the top ten. Unfortunately this is not the case. In the comparison based on the standard output of agriculture Hungary is the 13th biggest producer in the European Union. We must consider indeed, that every comparison using value can be distorted by the prices of the products. In this chapter we don't consider the factor of prices differences and we assume that the difference in the output is caused by inappropriate methods of organization. We are aware that this assumption simplifies extremely our approach but we don't want to examine the structure of grown products and the prices (Figs. 3, 4).

The next question is whether our assumption that Hungary is basically an agricultural country (because 78 % of the country is used for agro forestry activities) is true or false. If we are so lucky because of the land usage it must be represented in

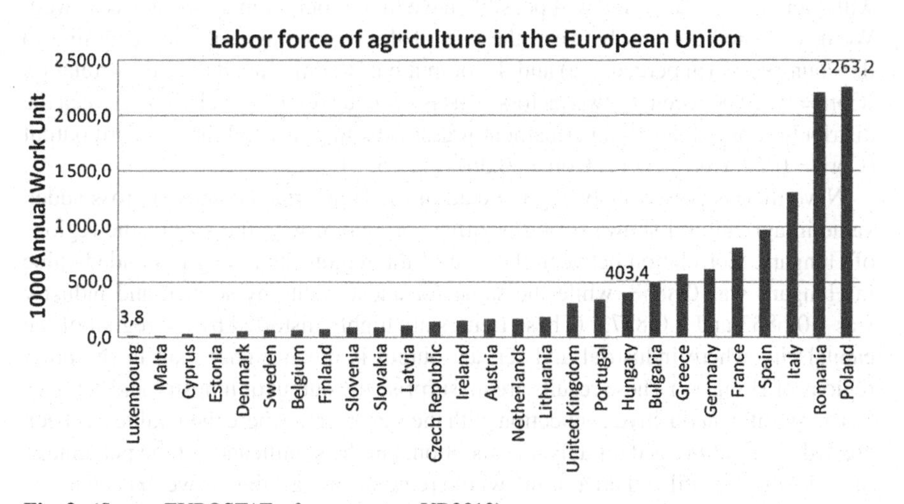

Fig. 3 (*Source* EUROSTAT—farm structure YB2013)

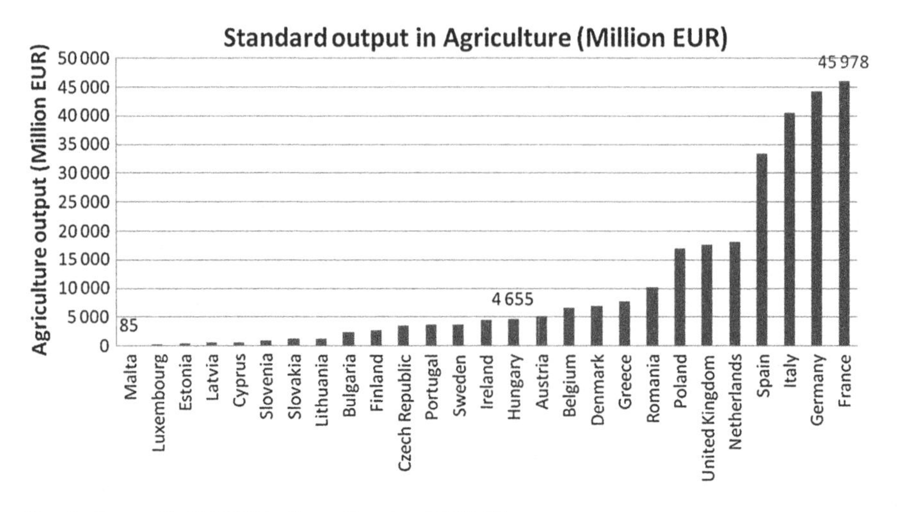

Fig. 4 (*Source* EUROSTAT—Farm Structure YB2013)

the economic indicators. The contribution of agriculture to the Hungarian GDP and to employment is about the EU-27 average. This shows that owning a great resource alone is not enough to have any kind of benefit from it. The world tendency is, of course, a decrease in the importance of agriculture in the developed countries.

In terms of the share of GDP or the employed workforce Hungary cannot be defined as an agricultural country even with so high share of the agriculturally utilized land.

According to our statistical analysis of the utilized agricultural area and the standard output of the EU-27 countries in 2000 and in 2010 we found high correlation (Corr = 0.86; Corr = 0.88) between the selected values. It is obvious that a country with more agricultural area will possible have higher output in agribusiness as well. We investigated the link between the size of the utilized agricultural according to the countryside (in percentage) and the output but we only found a weak correlation (Corr = 0.32) between the two factors. This is also true for the correlation between the distribution of arable land, permament grassland and crops and the standard output (Corr = 0.24; Corr = 0.11; Corr = 0.36) (Figs. 5, 6).

Nevertheless (when analyzing the data of the Hungarian economy) gross added value in agricultural shows strong positive correlation with the size of the regions of Hungary. Correlation between the size of the regions and the gross added value in Hungary was 0.8848, while the same for added value by service and industry was −0.6853 and −0.8173. (These figures are highly distorted by the region of the capital city where industrial and service output is much higher than in the other regions of Hungary) These results are no surprise the output of industry and services has no significant positive connection with the size of area where their value has been created. Agriculture is definitely an exception, but the significance of the percentage of agriculturally utilized area of the whole region is not as high as we expected.

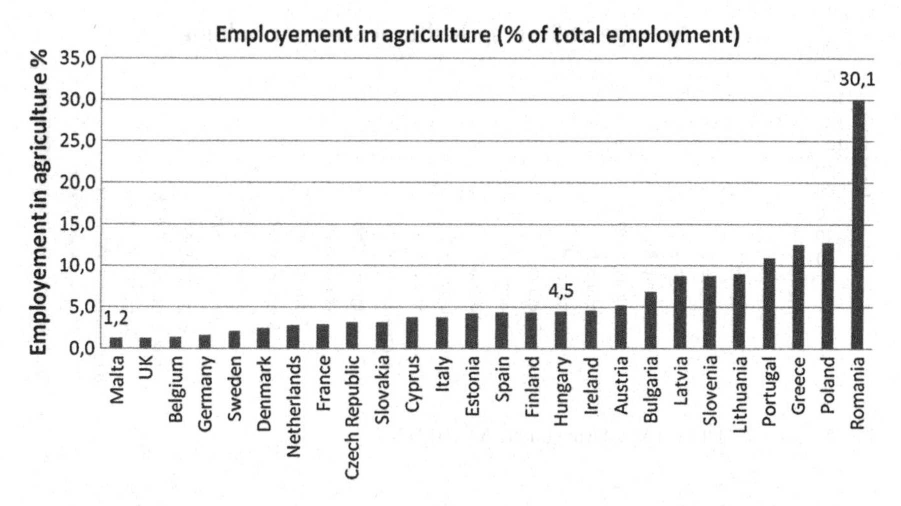

Fig. 5 (*Source* world bank database—world development indicators, 2010)

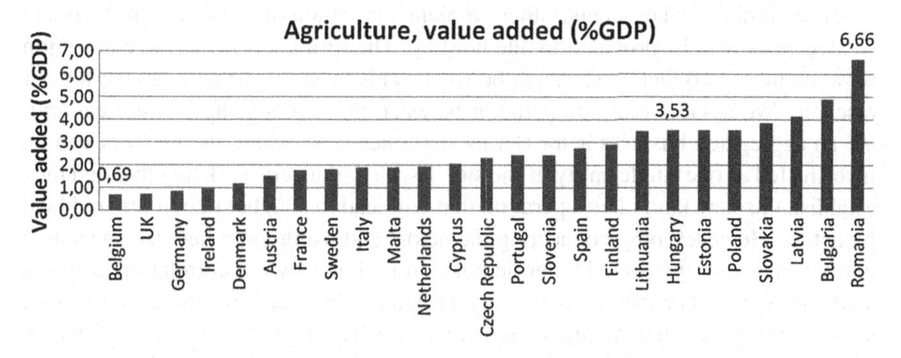

Fig. 6 (*Source* world bank database—world development indicators, 2012)

We also analyzed the correlation of the labor and the output of the agriculture at the EU-27 countries in 2010 and we found a mid-strong correlation between the two factors (Corr = 0.57). Unfortunately we cannot say that the distribution of full time workers in agriculture showed the same tendency. Between the percentage of full-time agricultural workers (compared with all the workers in the sector) and the output we found weak correlation (Corr = 0.27).

We also analyzed the correlation between the output and the machinery supply. Using the data of Gross value added of the agricultural industry—basic and producer prices and the number of tractors per 100 sq. km of arable land we found a correlation of 0.46 using the data of Hungary and 0.51 with the data of Germany. We found relevant data for the European Union between 2001 and 2005 but changes generated by the new members distorted the results.

The next step was calculating the correlation between the average utilized agricultural area per holding and the output. An agricultural holding is a single unit,

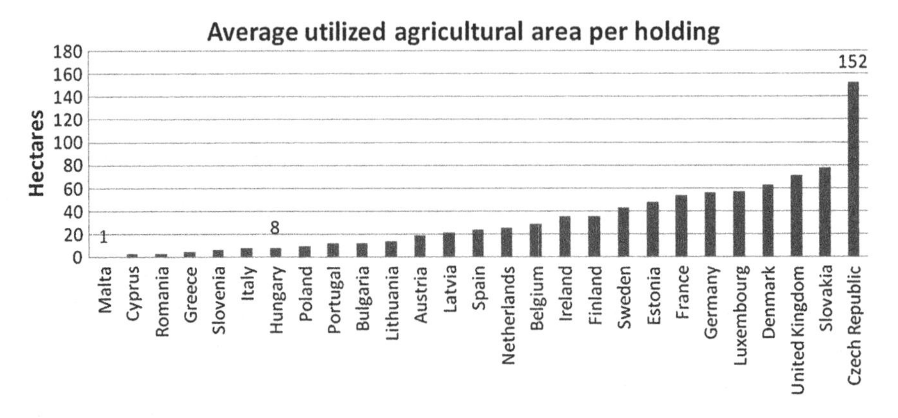

Fig. 7 (*Source* EUROSTAT—farm structre YB2013)

in both technical and economic terms, operating under a single management, which produces agricultural products. Other supplementary (non-agricultural) products and services may also be provided by the holding. The result is 0.15, so the connection between these two factors seems to be weak. This is an important message for our research. We assume that co-operation between the actors of agriculture can create an aggregated business actor (let us say a network) where the network can be investigated as one single entity. If the small-scale producers work together and they establish this new kind of co-operation that will lead to a higher rate of utilized land by actors. However, only connecting the actors and using more land by one single entity won't lead always to higher output as it is shown by our correlation analysis. There must be other factors as well which later help to achieve the economies of scale. We presume that the higher added value is based on the synergetic effects of the co-operations indirectly. This chapter does not focus on these problems.

Summarizing the results of the analysis we can assume that the output of agriculture has significant connection with the size of the utilized agricultural area (which is quite obvious) but there is no extra advantage if a country's land is mostly utilized by the agriculture. The number of workers and the machinery in the agricultural sector are also important factors, but they do not explain all the changes in the output (Fig. 7).

2.2 Structure

Another problem can be that the structure of the agricultural sector is fragmented. Most of the actors in this sector are small-scale (primary) producers, the basic elements of Hungary's agricultural system. There more than 400,000 primary producers around the country who do not have relevant co-operation with each other. Many of them are unable to market their own products because they are far from achieving

Table 1 Types of companies in the Hungarian agricultural sector (*Source* Hungarian central statistical office (HCSO))

Company/Year	2008	2009	2010	2011
Limited liability company	7214	7279	7684	8247
Public limited company	320	315	319	313
Limited partnerships	3923	3458	3201	3007
Cooperatives	1136	1004	960	891
Small-scale producer	355,431	376,305	388,440	400,368

Table 2 Distribution of agricultural companies by the number of employed (*Source* Hungarian central statistical office (HCSO))

Number of employees	2008	2009	2010	2011
500<	9	7	6	8
250–499	26	24	25	23
50–249	387	353	317	296
20–49	661	647	633	617
10–19	877	796	781	772
1–9	362,805	354,731	351,000	343,970

economies of scale, i.e. they do not grow enough to cover the costs of marketing, shipment, etc. Without co-operation or major investments they will not reach this level. To see the significance of this, we must know that the primary producers create 60 % of the value in the sector, but they are not working together so the effectiveness in insufficient. This could be a potential field of intervention. we believe that in network-type co-operations these producers can be supported and their products can be sold through a well managed supply chain. We will refer to this kind of solution as a horizontal network where the coordinator actor will manage the collection and merchandise of the products (Table 1).

If we are looking for a coordinator actor we should think about a large-scale company who can integrate the small producers and can enter the market with the aggregated supply. Mostly these large corporations are unwilling to co-operate (rather they take advantage of others) with the small ones. This can be an option for a horizontal co-operation where the small-scale producers are linked in a supply chain to the greater companies where their products can be marketed jointly (Table 2).

Another type of co-operation can be the integration of several actors of the sector of agriculture in a vertical co-operation. In the socialist era co-operatives were the forms of forced co-operation of the producers and till nowadays the expression is often connected with those communist ideas. Despite that the idea of an integrated farm is suitable for the agricultural works. This is because of the multiplication effect of the agriculture. Kovács [2] calculated that every 1 unit increase of the agricultural output creates a grow of 1.8 unit in Hungary's economic output. Vertical co-operations in a region in the field of food processing can have even better impacts.

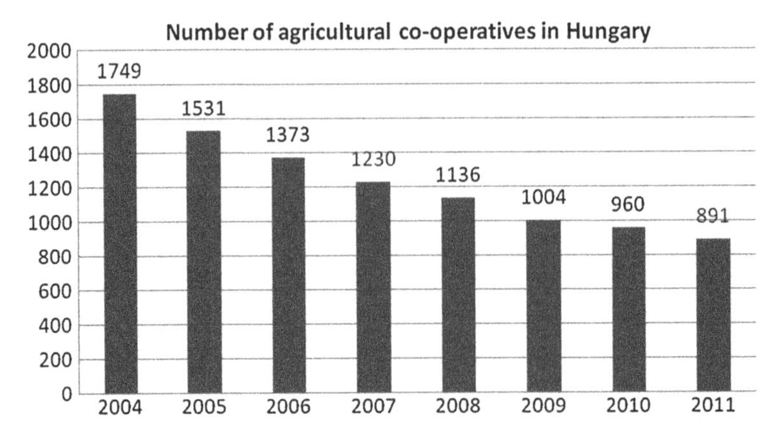

Fig. 8 (*Source* Hungarian central statistical office (HCSO))

Imagine that the consumption of Hungarian food products would increase with 15 %. This would generate 10,349 jobs in food industry, 4,751 in agriculture and 5,840 in the supplier network [2].

Unfortunately there was no such increase in food industry or in agriculture in Hungary. The number of agricultural holdings decreased with 125,000 between 2000 and 2010. In Hungary the degree of decrease was greater, more than 390,000 holdings disappeared. Simultaneously the average utilized agricultural area per holding increased in the EU-27 from 25.5 hectares to 31.9 hectares. This is about an increasement of 25 % while in 2010 a Hungarian agricultural holding used 72.5 % more land than in 2000. Although if we compare this data with the data of other European countries we can realize that an average Hungarian holding manages smaller territories than the average European. This is also a sign of the fragmentation of the sector.

Later on we decided to investigate the livestock breeding data as well (which is represented in the appendix) but we also experienced a steady decrease in the number of swine and cattle breeding. (Number of swines decreased with 1 million and cattles with 130 thousand between 2004 and 2011) (Fig. 8).

After facing all the facts mentioned in this chapter we decided to look forward to other parts of agriculture where co-operations can be defined and the links between the different actors can be interpreted easily. There is one thing common in every activity of the agriculture: waste is generated and it must be handled by following strict laws and regulations. So our next step was the analysis of the amount of waste in Hungary.

2.3 Selecting the Region

The biggest amount of waste created in Hungary in the year 2011 was coal ash in an amount of more than 1 million tones. We are not interested in all kinds of wastes so we searched for the specific types that are created in agriculture. The biggest amount

Table 3 Added value of agriculture in NUTS 2 regions of Hungary (million HUF)

Region/Year	2007	2008	2009	2010
Közép-Magyarország	975,589	86,182	59,968	66,018
Közép-Dunántúl	106,911	105,814	89,788	108,042
Nyugat-Dunántúl	115,763	106,565	90,081	104,517
Dél-Dunántúl	121,043	126,608	108,575	124,963
Észak-Magyarország	75,549	84,335	66,210	74,245
Észak-Alföld	179,371	189,639	158,495	178,937
Dél-Alföld	220,686	212,886	173,521	196,013

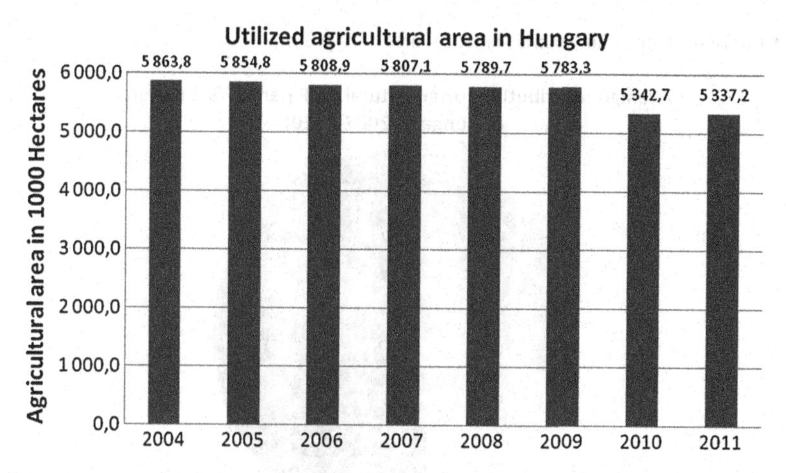

Fig. 9 (*Source* Hungarian central statistical office (HCSO))

of waste in agriculture is animal urine and manure (including spoiled straw). 59,640 tones of this type of waste were generated in Hungary in 2011. All together about 400,000 tons of waste is generated by agriculture annually which can be used in anaerobe fermenting rooms to generate biogas. If we would like to use this kind of primary material we must find the biggest biodegradable waste producers in Hungary (Table 3).

The biggest agricultural waste producer region was Észak-Alföld and Dél-Alföld (180,000 and 110,000 tons in 2010). Also these two region generates the highest added value in agriculture.

We made an average of the regional added value of agriculture in the last four years (Fig. 9) and we can see that about 45 % of the all agricultural output of Hungary was created in the mentioned two regions. We assumed that if these regions are the biggest agricultural producers than we must search for the companies in the regions to find a coordinator. Our search was based on the production of waste so first we examined the sub-categories of waste in these two regions (Figs. 10, 11).

In the region of Észak-Alföld the three biggest biodegradable waste producer operates in the field of dairy cattle breeding, pig breeding and poultry processing and

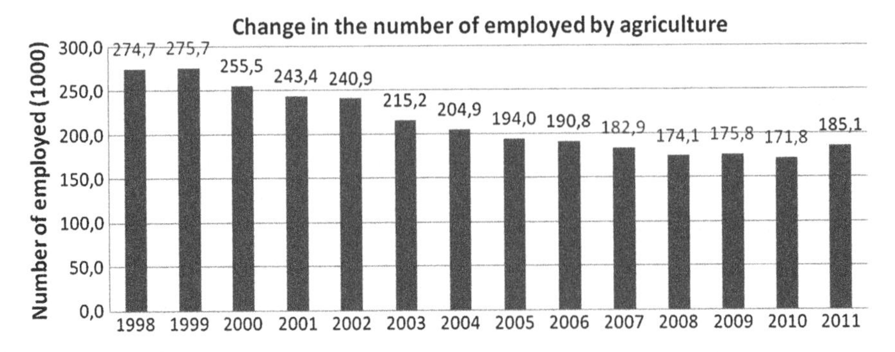

Fig. 10 (*Source* Hungarian central statistical office (HCSO))

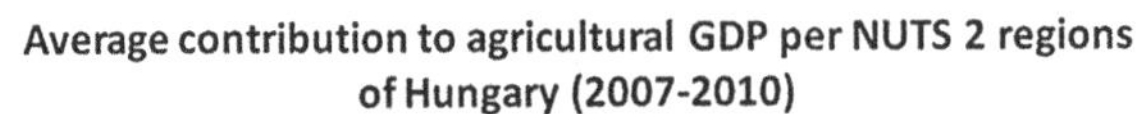

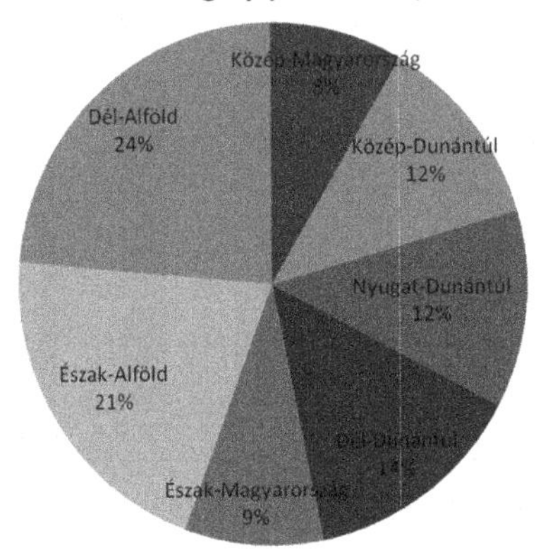

Fig. 11 (*Source* Hungarian central statistical office (HCSO))

preservation. The situation in Dél-Alföld is almost the same, the three biggest waste producers are the pig and poultry breeders and the meat processors .

We presume that those companies who produce copious amount of biodegradable waste would be interested in a way where they can benefit from their "useless" output rather than facing severe difficulties with the deposition of the waste.

From our point of view the solution can be establishing networks where actors work together to collect and reuse the biodegradable materials to create electricity. This idea has been arised when we got acquinted with the idea of integrated energy farms at Abony Mezőgazdasági Zrt. The main idea of the energy farm concept is full range recycling of the manure and other agricultural wastes within the farm. This is performed by anaerobic fermentation, using the biogas for electricity and heat production. This energy is used by the farm and any surplus can be traded. A biomass power plant is able use all other wastage to create energy. A short summary below shows the possible material which can be used in the biogas fermenting room and in the biomass plant:

- Plant-based cellulose of residues or livestock manure generated (mainly cattle, pigs, poultry, sheep, etc);
- Commercial wastes, spoiled food;
- Grape and wine by-products grape, shoots, cuttings;
- Organic (plant-based) oil production wastes (glycerol, oil-cake, etc.);
- Other organic wastes, such as trimmings from squares, parks, roadside banks;
- Sites of selective collection of waste (canteens, restaurants, supermarkets, etc.);
- Fruit and vegetable waste from processing plants (tomato peels, sugar beet cleaning wastes, fruit wastes, etc.).

The biogas is created during the process of anaerobic fermentation. The gas is used directly in the biogas power plant, where electricity is generated. The waste heat is used to heat the cattle barn. The electricity generated at the farm is consumed by the dairy coolers and by other equipment. This electricity is enough to meet the needs of 15 % of the farm's energy consumption. The waste heat is also used by the energy farm's bio-ethanol plant for the procedure. Barns, stalls and other buildings (foil heat, greenhouses, etc.) can be also heated by this waste heat. The slurry fermented in the biogas plant can be divided using a separator. The liquid part contains about 3 % nitrogen so it is suitable for irrigation. The solid part can be used again in the fermenting room. This process does not lead to further CO2 emissions, so it does not increase the greenhouse effect.

Back to our primary objective (regional development using agricultural potentials) we assume that fermenting rooms and biogas power plants should be deployed in the optimal regions to reuse the waste material of agriculture. This can be an option to find a new field of agrobusiness. The task of the deployment is complex and influenced by several factors. We suggest that these energy farms must be created where there is already agricultural activity and where the companies can consume their own waste to ensure their own needs of electricity. This operation can be also performed in a network type co-operation. All cooperation among economic actors can be defined as a kind of network. From this point of view the whole global economy is one enormous network where smaller networks are competing. There is no general agreement on a definition for the term of business network or business cluster. Often a business cluster is "a geographic concentration of interconnected businesses, suppliers, and associated institutions in a particular field. Clusters are considered to increase the productivity with which companies can compete, nationally and globally" [3]. For this article let us define a business network as a conglomeration of cooperating producers and companies where they jointly create value needed by the customers. Network type cooperation is one of the best solutions for high flexibility, regional development, and to ensure competitiveness of the members. Network cooperation is usually more beneficial for the members than individual work. The benefits usually come from the synergetic cooperation of the members. There are different co-operational styles and forms for companies but in this chapter we only present one example of the horizontal network. A horizontal network in agriculture can be co-operation between the smaller waste producers and a large-scale company, the coordinator. The coordinator's role is to create a link between the small actors and the energy farm. The coordinator's work is basically logistical. Beyond the collection and distribution of waste the coordinator has to create the information system that can handle the demands of the customers and that can also create a supply chain structure of the waste producers.

3 Logistics in Energy Farm Networks

Energy farm based network are a special kind of hub-and-spoke networks which always have a central unit performing the activities, with all the other members also involved. Nevertheless, in logistics networks several other units can join in the function of waste collection, with several landfill sites. Biodegradable waste is also generated in scattered locations in a region. A substantial amount of waste will usually not occur particularly in one certain location, so different subunits of collection must be deployed. The basic network elements in this case are related to agricultural production and production lines, which generate biodegradable waste that can be used in the fermentation process or ethanol production. Logistic service providers such as warehousing companies or forwarding agencies are also involved in the network's activity. Internal and external processes are also considered by the energy farm. Not only the routes of collection are important but also the deployment of the energy farm units (such as the power plant or the storage) must be examined. The core of the network is always the energy farm (situated as a hub). This hub creates the links with the other spokes, in this example with the farms, slaughterhouses, etc.

Energy farm networks can lead to several advantages for the network members:

- Fermenting room is used optimally,
- Better utilization of transport vehicles,
- Smaller quantities and more frequent deliveries of raw materials,
- Reduction of the environmental impact of the organization.

A network-based logistics system operates with the mathematical models of route optimization and core modeling procedures. For this task we can use the support of different software. Basically the general logistical tasks of the network according to Cselényi and Illés [1] are:

- Defining the number and location of the network members who are involved in logistical processes (in this case this is the energy farm itself and the different locations with waste),
- Defining the resource units, capacity and characteristics of the transportation (manure and some other wastes are considered hazardous waste),
- Establishing the network organizational structure,
- Developing strategies for route management,
- Scheduling logistical activities,
- Tracking the material flow in the network,
- Establishing a proper information system for designing and operating the network's logistical processes.

Regarding the fact that the primary hub is the energy farm and the logistical processes can be identified as a special form of supply chain, we must characterize the activities of the supply chain and the material flow of the chain. Material flow systems can be described by the following characteristics according to Cselényi and Illés [1]:

- Method of delivery (delivery is possible only by road),
- Type of goods delivered (hazardous waste),
- Means of transport (special closed space, self-loading refuse collection truck),
- Location of entry characteristics (a point to accept the incoming vehicles),
- Characteristics of the receiving surface,
- Location of storage space,
- Suppliers, loading equipment connection,
- Vehicles carrying supplies to exit (point of entry).

Recycling and waste management is a specific area of logistics. The processes of waste management, especially biodegradable wastes (such as manure) are legally defined. Beside the technical questions, all regulations must be considered as well when designing the routes or selecting the transport vehicles. Recycling and waste management considers a wide range of diverse waste but in this chapter only biodegradable wastes were considered.

When the specific method of waste management is selected the following questions concerning the development of logistic systems should be answered according to Cselényi and Illés [1]:

- The number of stages of collection (one-, two- or three-stage sampling),
- The collection centers,
- The collection and processing of material and information flow relations,
- Collected homogeneity of the product (for easier handling of specialised equipment),
- Specific requirements of transport application due to hazardous waste,
- The issue of network management.

The number of stages of collection depends on the amount of waste and on the location of the places were waste is generated. If the waste is generated densely in the area one-step-collection can be performed. Otherwise several centers must be established to ensure the optimal collection service in the area. This problem must be solved by the regions mentioned in Sect. 2.3 by the biggest waste producer companies.

If one of the waste producer companies would like to act as a coordinator in the network they should consider all the logistic tasks and processes of the operation. These operations can be evaluated by the following parameters of a logistics network according to Cselényi and Illés [1]:

- Collection system design, equipment selection,
- Storage space design,
- Choice of instruments forming unit load,

- Storage space, storage mode,
- Selecting means of transport,
- The acquisition, development of local landfills,
- Economic incentive system to facilitate the collection,
- Public awareness, (a part of municipal waste can also be connected to the system),
- The collection of system management features,
- Scheduling and management of collection services,
- Disposition of the assets of conveyor systems,
- Select applicable management system,
- Management control functions, design,
- Legal and administrative systems selection,
- Operation of complex simulation tests.

In the logistic system of waste disposition several companies can work together to ensure the optimal functioning of the hub-and-spoke network.

4 Conclusion

The differences in regional economic development (acknowledging the endogenous theory of growth) do not disappear automatically. Business networks that aim to enhance regional development should be only based on bottom-up initiations. These initiations can be supported by the tools and methods of network establishment using logistical functions. In every country it is reasonable to use biodegradable waste to provide electricity to the countryside and farms. Due to the complexity of the agricultural works, several links can be established between the members. One of the links could be the waste management of the members. This primary connection can be the foundation of a network operating in a hub-and-spoke form. An energy farm could operate as a hub in the network. Optimal logistic processes are inevitable for a long-term economic operation in a network. Logistic processes can be defined within the energy farm and among the members. This leads us to a complex view of internal and external logistical processes in the whole network. The quality of external links can help the network to evolve and to establish more interrelations between the members. The external processes are basically the core search process and route management. Both of these processes are complex and can be divided into several activities which should be performed by the logistics management company involved in the network establishment.

Appendix

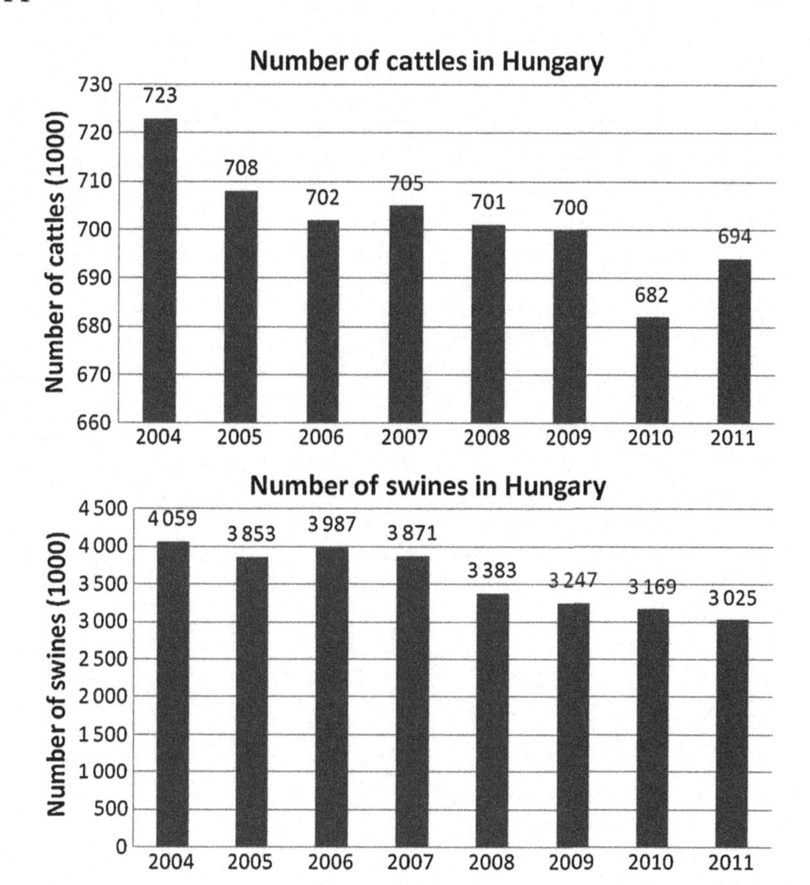

Acknowledgments The described work was carried out as part of the TÁMOP-4.2.2/B-10/1-2010-0008 project in the framework of the New Hungarian Development Plan. The realization of this project is supported by the European Union, co-financed by the European Social Fund.

References

1. Cselényi, J., Illés, B.: Logisztikai rendszerek I. ("Logistics systems I."), Miskolci Egyetemi Kiadó (2009)
2. Kovács, G.: A mezőgazdaság nemzetgazdasági jelentősége ("The importance of agriculture in the national economy") Gazdálkodás **5**, 466–479 (2010)
3. http://epp.eurostat.ec.europa.eu/statistics_explained/images/d/d0/Farm_structure_YB2013.xls
4. http://data.worldbank.org/data-catalog/world-development-indicators

Printed in the United States
By Bookmasters